Is That a Big Number?

IS THAT A BIG NUMBER?

ANDREW C. A. ELLIOTT

When you can measure what you are
speaking about, and express it in numbers,
you know something about it.
Lord Kelvin

OXFORD
UNIVERSITY PRESS

OXFORD
UNIVERSITY PRESS

Great Clarendon Street, Oxford, OX2 6DP,
United Kingdom

Oxford University Press is a department of the University of Oxford.
It furthers the University's objective of excellence in research, scholarship,
and education by publishing worldwide. Oxford is a registered trade mark of
Oxford University Press in the UK and in certain other countries

© Andrew C. A. Elliott 2018

The moral rights of the author have been asserted

First Edition published in 2018

Impression: 1

Published in the United States of America by Oxford University Press
198 Madison Avenue, New York, NY 10016, United States of America

British Library Cataloguing in Publication Data
Data available

Library of Congress Control Number: 2017962198

ISBN 978-0-19-882122-9

Printed and bound by
CPI Group (UK) Ltd, Croydon, CR0 4YY

BEWILDERED BY BIG NUMBERS?

There are 7.6 billion people in the world.
Is that a big number?

Life first emerged on Earth 4 billion years ago.
Is that a big number?

There are 1.55 million wildebeest in the wild.
Is that a big number?

A US Navy Supercarrier costs $10.4 billion.
Is that a big number?

A high-jumper clears 2.4 metres.
Is that a big number?

For my sons, Ben and Alex

CONTENTS

Introduction

Numbers Count

Without numbers, we can understand nothing and know nothing.   **Philolaus**

Which of these is the most numerous?
- ☐ Number of Boeing 747s built (up to 2016)
- ☐ Population of the Falkland Islands
- ☐ Number of grains of sugar in a teaspoon
- ☐ Number of satellites orbiting the Earth (in 2015)

At the start of each main chapter of this book, you will find small quizzes like the one above. The solutions are at the back of the book.

Introduction

In around 200 BCE, Eratosthenes calculated the size of the Earth. He was a mathematician, a geographer and a librarian, the third chief librarian of the famed Library of Alexandria. He knew that in Syene, some 840 km away, at midday on the summer solstice, a beam of sunlight would shine straight down a well shaft, the Sun being directly overhead. At the same time, in Alexandria, Eratosthenes measured the angle between the Sun and the zenith. The angle he measured was approximately $^1/_{50}$ of a circle. Using a little geometry (which, after all, means 'Earth-measuring'), he made a calculation of the world's circumference. His answer was accurate to within 15% of the true value.

Some 17 centuries later, when Columbus set out across the Atlantic, he disregarded Eratosthenes's calculation and relied instead on a map by Toscanelli, also a mathematician and a geographer, from Florence. Toscanelli had met with Niccolo Conti, the first Italian merchant to return from the Far East after Marco Polo's journey, and added this first-hand account to the knowledge he'd acquired in his studies. But Toscanelli had mistaken the size of Asia in his calculations, and Columbus ended up using a map that underestimated the size of the Earth by 25%. When he made landfall in America, he thought he'd reached Asia. Instead it was, to Europeans, a 'new world', something Columbus never acknowledged. Numbers are important: getting them wrong has consequences.

Numbers give size and shape to our world in all sorts of ways, and we rely on them to inform the decisions we make. It's easy though, when numbers become too large, to become numbed by sheer scale.

This is not a book of mind-boggling number facts, or stupefying statistics. This book is all about finding a path through a wilderness of big numbers that we don't grasp as well as we could. It's about mapping the landscape of numbers, learning how to recognise safe ground and how to spot boggy terrain. It's about providing landmarks to find our bearings, and keep us safely on course.

The world is chaotic

Recently, a particular image has come repeatedly to my mind. The image is of a body of water, a vast lake or river. The surface of this water is a jumble of flotsam: human litter, leaves, seeds and pollens, iridescent contours of oily pollution. The wind blows across the water and whips up waves chaotically. There are whirlpools, some tiny, some vast. On the shore, there are inlets and backwaters.

Look at a small portion of this vast watery canvas, and you will see the surface layer of detritus moving now one way, now another. The wind-whipped waves give an illusion of movement in one direction, but look closely and you'll see the leaves and cigarette butts drifting slowly in counter-movement, only to become caught in a gyre, and return to their starting places.

But this is the question: is the water as a whole flowing in any particular direction? And if so, in which direction?

This recurring image comes to me when trying to make sense of events in the world recently. Tragedies seem to unfold every day, every week, every month. Refugees make their way from unspeakable horrors, undertaking hazardous journeys, to futures that are at best uncertain. Demagogues and extremists of all stripes whip up resentment, hatred, and violence.

And yet, we (some of us, at least) live in a world of material luxury and life chances unthinkable to our parents. We know also that living standards have risen, not just in the developed world, but, taken broadly, across the globe. So, are things getting better or are they getting worse?

One way to gain some perspective is to try for a wider view. Instead of looking at a small patch of water from close up, climb the tallest tree you can. Look down from a height where the detail of the froth and the flotsam becomes lost. Look for the tell-tale colour variations that reveal where the deep water is flowing.

An understanding of the numbers in the news is essential to the construction of a reliable worldview that can guide us in knowing what really matters and what we can dismiss as froth. Numeracy alone is not sufficient, but it is necessary. At its best, a numerate worldview is testable (capable of refutation) and contestable (capable of being argued against). It is inherently self-challenging, since inconsistencies readily reveal themselves.

Scientists and engineers depend on a numerical understanding of their subjects. This underpins the stable and coherent models on which they base their work. Their efforts repeatedly test the correctness of their models by applying them in practical situations. Consequences tend to be immediately apparent. This is, for practical purposes, a working definition of truth—or at least a partial version of the truth of some aspects of the world.

To build our own numerical worldview, we don't all need the fluency with mathematics and numbers that engineers have. For most of us, everyday numeracy is enough. It's often enough to ask the right questions: 'Is this more, or is this less?' and 'Does that amount to a big number?'

Numbers count

Without numbers, our modern world could never have come into being. Numerical thinking has deep roots.

The first writing was done in the service of accounting: the Sumerians would encase small tokens representing the goods they were shipping, in a clay envelope, a *bulla*, as documentation, a manifest. In time, it became convenient to make imprints of the tokens on the outside of the envelope, saving the need to break it open. Later still, it became clear that the imprint alone would suffice: that the image could be drawn, and then reduced to a symbol, a glyph. There was no need for multiple copies of the same token: a different symbol could be used represent the quantity of items.

Numbers came early to the process of city-building. A small band of hunter-gatherers would have had a limited need for numbers beyond the most basic counting and sharing. But cities demand more complex calculations. Numbers are essential to building, trading, governing. The tablets found at Knossos in Crete, when deciphered from the script called Linear B, turned out to be government records. There is always administration; it is central to an ordered state. And numbers are central to administration.

From the start of writing, from the start of recorded civilisation, numbers have been used. The first verses of the Bible tell a story of counting: the counting of the days of creation. Later, the Bible details with precision the size of the Ark of the Covenant (and indeed, of Noah's Ark). The school of philosophers associated with Pythagoras declared that all was number, and all things came from numbers. Numbers have always counted.

Developing number sense

I use the term 'number sense'[1] to refer to our ability to comprehend numbers in an intuitive way, readily available to our thought processes, without relying on additional mental processing. In his book *Stranger in a Strange Land*, science fiction writer Robert Heinlein used the term 'grok' to mean 'to have a deep and intuitive understanding of', and that's close to what I mean here.

We learn to 'grok' numbers in the single-digit range before we reach school. For most of us, fluency with these small numbers becomes very deeply ingrained, to the extent that we scarcely recognise that saying 'I came fifth in the sack race' involves a numerical skill.

In the early years of school, our range increases, and we become very familiar with numbers up to the hundreds, and through the rest of our education and in adult life we extend this reach by perhaps a factor of 10, to the point where most people have a degree of comfort with numbers up to a thousand or so. Those who study the sciences will learn to manipulate and manage numbers of much greater magnitude, but such skill at manipulation is not necessarily the same as having a 'feel' for the numbers themselves.

[1] I've borrowed the term from the wonderful book *The Number Sense: How the Mind Creates Mathematics*, by Stanislas Dehaene.

When we encounter bigger numbers, as we do especially in regard to public life at the national (or international) level, we are less secure. Whether we're faced with immigrant numbers, national budgets or deficits, costs of space programmes, health services, or defence budgets, we lack the ability to easily put these into context. In large part, this is simply because we cannot fully grasp the large numbers themselves.

Five ways to think about big numbers

This book presents various mental strategies for thinking about big numbers grouped under five main headings, all in the service of an overall guiding principle.

That guiding principle is **cross-comparison**. The best way of understanding what a big number means is to contextualise it, to find meaningful comparisons and contrasts, to connect the big number to other, known measurements. There are all sorts of ways to do this, but in this book I present five main approaches, each of which reinforces the others, all of which contribute to building that network of number facts which help create a numerate worldview:

- **Landmark numbers**: Using memorable numbers to make instant comparisons or as yardsticks.
- **Visualisation**: Using imagination to picture the number in a context that allows mental comparisons.
- **Divide and conquer**: Breaking the big number down into smaller parts and working with the parts.
- **Rates and ratios**: Bringing numbers down to size, by expressing them as a proportion of some base.
- **Log scales**: Dealing with numbers of widely different scales by measuring proportionate variation and not absolute difference.

Each of these approaches will receive a turn in the spotlight at an appropriate point in the book, but they'll be in use throughout the text. Watch out for them!

Everything connects

Just as in a detective thriller, one clue on its own may be hard to interpret, so one number fact in isolation may not tell much of a story. But when the number 'facts' start connecting, the mystery begins to unravel, and a version of the truth emerges. For this project, I've been building a collection of juicy number facts and storing them away in a database. Any one of these number facts on its own may be unremarkable, but when they are assembled as a collection, you start to see the connections and then to actively seek others. You learn that the first printing press is around five times as old as the first transatlantic wireless message (577 years versus 116 years). You learn that a pint and a pound of weight and a pound of money are all connected in words and ideas (and connected again to the *lira* and the *livre* and to Libra, the constellation in the Zodiac, and star sign). You see the imbalance between growing human population numbers and the shrinking numbers of animals in the wild.

Connecting one fact to another, you begin to knot them into a net, to trap the new number facts or assertions that come your way. Thus equipped, you're better able to assess their worth: treasure or trash? So much of the value of these numbers lies in their contexts, in their combinations and contrasts.

Numerical serendipity

This is not a dry list of isolated facts. The IsThatABigNumber.com website has code that trawls its database in search of unusual pairings of number facts arising from random juxtapositions.

Some examples:

- St Paul's cathedral in London is almost exactly 100 times the height of R2D2, the droid in *Star Wars*.
- Archimedes was born about 4 times as long ago as Leonardo da Vinci (2300 years versus 565 years).
- The Great Wall of China is 10,000 times as long as Tiananmen Square.
- There are about 1000 books in the British Library for every polar bear alive.

For your numerical delight

There's a subtle and geeky joy in knowing stuff and being able to work out stuff for yourself based on what you know. The landmark numbers and rules of thumb you find here will show you a way of looking at the world and knowing more about everything you see. The precepts and the examples in this book will help you to know a bit more about, and to assess more clearly, the numbers in the world around us, and what they mean. But much of the book is simply a celebration of numeracy—counting and measurement—and the deep way in which numbers are embedded in and enrich our understanding of our world and our lives.

Random alignments

Scattered through this book, you will find panels like the one below. These are pairings of numbers that happen to have a neat ratio between them. The exactness of the match is always within 2%. There is no reason for these connections to be meaningful (but they do show how easy it can be for conspiracy theorists to find coincidences around which speculation can be woven). And even though the pairings have no more meaning than when there is a conjunction of Mars and Jupiter in the night sky, the comparisons often manage to bring a perspective to the numbers they refer to, and to help develop a sense of number or magnitude.

Did you know? (Spurious number coincidences)

Length of a classic **Ford Mustang** (4.61 metres) is
about as long as a **great white shark** (4.6 metres)

Age of earliest stone **hand axe** (2.6 million years) is
500 × age of earliest evidence of **writing** (5200 years)

Mass of **Mercury** (330 sextillion kilograms) is
25 × mass of **Pluto** (13.11 sextillion kilograms)

Highest ascent by a **propeller-driven aircraft** (29.52 km) is
5 × height of **Mount Kilimanjaro** (5.89 km)

Length of the **River Thames** (386 km) is
2 × length of the **Suez Canal** (193.3 km)

Time since the birth of **al-Khwarizmi** (1240 years) is
4 × time since the birth of **Euler** (309 years)

The First Technique: Landmark Numbers

When You're Lost, Look for Landmarks

I grew up in South Africa and moved to Britain as an adult. I had very limited exposure to and hence little knowledge of British history. Then, a few years after my arrival, a friend recommended that I read Samuel Pepys's Diary. Not my usual reading choice, but I trusted his judgement, and I found to my delight that it was lively, funny, touching, and very informative about the world that Pepys lived in. I pass on that recommendation at every opportunity and I do so now: you should read *The Diary of Samuel Pepys*!

His first diary starts on the first day of 1660, a year that was to be very important in British history: it's the year that King Charles II returned to Britain from exile—the Restoration of the Monarchy. Pepys had a junior role in the logistics of the return of the king (he ended up looking after the king's dog on the boat that shuttled between ship and shore).

He notes:

I went, and Mr. Mansell, and one of the King's footmen, with a dog that the King loved, (which shit in the boat, which made us laugh, and me think that a King and all that belong to him are but just as others are) . . .

Now that's a memorable passage, but never mind the king's dog: what stuck with me was that single date, 1660. That date has become, for me, a **landmark** date.

And that has transformed my understanding of history. These days, whenever I read of a historical incident, or read about famous people in history, that one date is my reference point. Shakespeare—his dates are before 1660;

Newton—born before that date (but his career is after); James Watt—after that date. That single date has done remarkable duty: giving me a context to locate so many other events in time.

And this is true not only of dates. Here's a small selection of landmark numbers that appeal to me, and might appeal to you too. They're all approximate: they're 'more or less' correct, but are round enough to remember, and accurate enough to allow you to answer the question: Is that a big number?

- Number of people in the world—7+ billion and growing
- Number of people in the UK—60+ million and growing
- Number of people in the USA—300 million
- Number of people in China—1.3+ billion
- Number of people in India—1.3 billion and growing
- UK GDP—2½ to 3 trillion USD[2]
- UK National Budget—1+ trillion USD
- USA GDP—18 trillion USD
- USA National Budget—6 trillion USD
- Length of a bed[3]—2 metres
- Length of a football field—100 metres
- Distance walked in an hour—5 km
- Length of the equator[4]—40,000 km
- Height of Mount Everest—9 km
- Depth of the Marianas Trench—11 km
- Teaspoon—5 ml; Tablespoon—15 ml
- Wineglass—125 ml; Cup—250 ml
- −40 °C = −40 °F (same temperature, same number)
- 10 °C = 50 °F (a cool day)
- 40 °C = 104 °F (a bad fever)
- A human generation—25 years
- Fall of the Roman Empire—500 CE[5]
- Start of writing and written history—5000 years ago

---

[2] Using US dollars to measure the UK economy helps with comparability.

[3] Don't laugh, I always judge the size of a room mentally by thinking about how it would accommodate a bed. So, for example, a 2 m by 4 m room would be pretty small—a bed would take up all of the short side.

[4] Length of a meridian—pole to pole = 20,000 km. So, pole to equator = 10,000 km, which is the basis of the original definition of the metre.

[5] I've used BCE and CE in this book rather than BC and AD, in line with common current practice.

- Modern humans spread from Africa—50,000 years ago
- Extinction of dinosaurs—66 million years ago
- Samuel Pepys starts his diary—1660 CE

You'll find many more of these landmark numbers throughout this book. I'm not saying that you should learn them, but the ones that most appeal to you will resonate with you. Some will stick, and you'll find them useful in all sorts of important ways, whether it's making rapid judgements of the 'more or less' kind, or as a crude yardstick for establishing context, or just knowing when someone is trying to pull the wool over your eyes.

PART 1

Counting Numbers

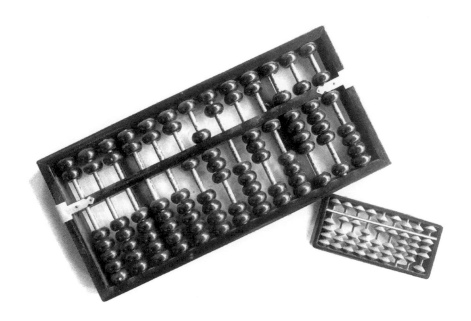

What Counts?

How We Get from 1, 2, 3 to 'How Many Fish in the Sea?'

If you think dogs can't count, try putting three dog biscuits in your pocket and then giving Fido only two of them.   **Phil Pastoret**

Which of these is the most numerous?
- ☐ Number of aircraft carriers in the world
- ☐ Number of skyscrapers in New York City
- ☐ Estimated population of Sumatran rhino
- ☐ Number of bones in the human body

Counting up

— There are 200 million people in Brazil. *Is that a big number?*

— There are 564,000 words in Tolstoy's *War and Peace*. *Is that a big number?*

— 438,000 people in the world died of malaria in 2015. *Is that a big number?*

These are all counts of things, but each has been counted in a different way. And before we can start giving sensible answers to these questions, we need to think a little about counting itself, that most basic of numerical skills.

What does it mean to count?

Let's check the etymology:

count (v.)
mid-14c., from Old French **conter** *'add up,' but also 'tell a story,'*
from Latin **computare**

https://www.etymonline.com

This book is about numbers, and to tell a story about numbers I must begin with counting. However big the numbers might become, it all starts with counting.

As children, we learn early on how to identify and label abstract qualities.[1] Though we may not realise it, those bright picture books are teaching the art of abstract thinking. The redness of an apple is like the redness of a riding hood, the roundness of the same apple is the roundness of the full Moon. And when five apples are placed in the basket to take to grandmother, we learn about number.

These are the counting numbers, the positive integers: five apples, five pears. From this will grow the abstract notion of the number five. The fiveness becomes a thing of its own, independent of the actual five apples, independent of any apples at all. But what is 'fiveness'? You can't touch it or taste it or see it or hear it, but you can name it, discuss it, remember it. 'Number' is one of the first abstractions we learn in the world. It's a Platonic ideal, and yet a three-year-old can grasp it.

We learn that these counting numbers underpin other concepts such as 'more than' and 'less than'. The person who comes fourth in the race is labelled by the number '4' but is not 4 people. Above all, we become comfortable with the notion that aspects of the physical world can be represented abstractly.

This notion lies at the heart of all scientific thinking, all 'theorising'. Using this approach, we can make predictions about the universe that are precise and accurate enough to create machines that fly, to bounce light beams off the moon, to send a spacecraft to Pluto.

And the first steps to this are taken when toddlers learn to count.

---

[1] When I say that we learn these things, I don't mean that we sit and study them in these terms. I mean that everyday curiosity, experience, and childish learning smuggles these profound ideas into our minds. But for all the childish garb in which they are clothed, the ideas are no less profound.

Counting (like) crows

There's a story about how crows can count. A landowner has an old tower on his estate. An annoying crow roosts in that tower, and he wants to kill it. But when he goes to the tower with his shotgun, the crow flies away. When he walks away from the tower, the crow flies back. So he has a plan: he will go to the tower with a companion, and let the companion leave, but he will remain. He reasons that when the crow sees his friend leaving, it will fly back and he will shoot it.

But the crow is not fooled: only when both people have left the tower does it return. So the man comes with two companions, making three of them in total, and two of them leave. The crow is still not fooled, and stays until all three have cleared off. He increases his party to four, and three of them leave. Still no success. Five arrive and four leave. But still the crow is too smart to be fooled. Finally, he has enlisted five friends, so that a party of six arrives, and five are seen to leave. At last the crow loses track of the number and returns, to be shot. So the point of the story is that crows can count, but only as far as five.

The story has a grain of truth to it: crows are indeed remarkably intelligent birds, and have been much used in experiments testing animal intelligence. They do have a sense of number. In one controlled experiment, a crow was able to choose which of several similar containers contained food, by recognising the number of dots marked on the lid.

Other animals, too, show similar levels of number-recognising and counting ability. These capabilities are typically effective only at very small numbers, and become unreliable when the numbers get beyond five.

Humans share this primitive number sense with other animals. In his book *The Number Sense: How the Mind Creates Mathematics*, Dehaene distinguishes two aspects of this innate numerical capability.

The first is called 'subitising' (from the Latin *subitus* meaning 'sudden'). This hard-wired ability allows us to directly and more or less instantly perceive quantities from 1 to around 4, without counting. Throw a small number of beans on the table, and you'll be able to see how many there are instantly: for bigger numbers, your performance becomes erratic. Throw 11 beans, and you'll have to count them.[2] Without the ability to count in a systematic way, even 10 is a big number, beyond direct perception. The first three Roman numerals are just vertical marks, but the Romans knew that rapid reading of numbers can't

---

[2] Certain people referred to as 'savants' appear to be able to instantly perceive numbers that are far greater than this. Current thinking suggests that such savants may have mental pathways that allow them to directly access perceptions at a less-processed level than the rest of us.

rely on counting scratches. From around 'IV', and certainly by 'V', a distinct symbol is needed.

Strictly, subitising refers to grasping a number directly without the benefit of pattern recognition. Some extend the meaning of the term to include those numbers we recognise immediately because of the patterns they form—for example, the sixes on dice, or the pattern of pips on a domino or a playing card, or the formation of players on a football field. If the objects we see fall into patterns, by design or by chance, we can, as a learned skill, identify their number quickly. But this is a learned skill, not an inbuilt capacity, and is reliant on the counted items forming patterns.

Beyond subitising, there is evidence of a second inherent numeracy skill, an ability to perceive approximate numbers, also present in other animal species. Dehaene writes:

> …scientists, when they describe perception of numerical quantities, speak of 'numerosity' or 'numerousness' rather than number. [This] enables animals to estimate how numerous some events are, but does not allow them to compute their exact number. The animal mind can retain only fuzzy numbers.

So, when comparing two groups containing, say, 80 items and 85 items, we see them as having 'more or less' the same number. The imprecision inherent in this ability is reckoned to be around 15–20%, so that while 90 might look approximately the same as 80, we'd be more likely to spot that 100 items looked noticeably different to 80. It is this skill that the crafty crow was displaying and which in the end let it down when the numbers became too big.

Another approach that we often use for rapidly counting groups of modest size, perhaps between 5 and 20 items, is to break a larger group into subgroups. It's relatively easily to perceive a group of items as a combination of smaller groups. Our minds can then use subitising or pattern-recognition skills to count each of those smaller groups, and then use small-number arithmetic to get to a total very rapidly.

This ability is interesting because it shows some features that we will see again and again when dealing with much bigger numbers. It's a three-stage approach, bringing together multiple skills: mentally splitting the bigger group, determining the counts of each of the smaller groups, and then recombining those sub-counts to get the overall total.[3] There is an algorithm at work here, a

---

[3] This is remarkably similar to the dominant principle used by information technologists in Data Science, which is known as 'Map-Reduce'. First you break a problem into parts, then you 'map' a function (in this case 'counting') to the parts, and then finally you reduce the partial results to a single overall result.

specific series of steps that our minds carry out in order to get to the right result, and it's one of the family of mental strategies that I call **divide and conquer**. We'll see this again when we look at how we cope with much bigger numbers.

These two abilities—subitising and the approximate estimation of numerosity—may be the only inherent number perceptions we have. All our precise numeracy skills beyond a count of 4 are based on other strategies: procedural combinations of mental skills and external aids.

The calculus of counting sheep

To get beyond the limits of subitising and small-number pattern recognition, we have to use strategies involving process and memory: and the first of these are systematic counting strategies. Let me introduce a little *calculus*.

Calculus? Relax—not that kind of calculus. 'Calculus' is the Latin word for pebble, and its mathematical sense comes from the ancient use of pebbles as calculation and counting aids.[4] Picture a shepherd sitting on the hillside, and in a pouch or a pocket she has as many pebbles as she has sheep in her care. When the sheep are gathered in at night, for each sheep that enters the fold, she transfers a pebble from one pocket to another. Only when she has no pebbles left in the first pocket is she sure that the sheep are all safely home.

So she has 'counted' the sheep, but to do so she has used a technique based on a process (the matching of one sheep to one pebble) and on memory (in this case, not her own memory, but the pebbles in her pockets). No doubt she could rely on her own memory, but she prefers the external record-keeping device because chasing sheep can get distracting.

Counting is singing

But sometimes the shepherd does use her own memory, such as when she is counting the balls of wool she has spun. She will still carry out a one-for-one matching technique, but this time she will match the balls of wool against a memorised sequence, the number names of her language: 'one, two, three' and so on.

---

[4] It's been speculated that the oval shape we use as the numeral for zero has its origins in the imprint left when a round pebble is removed from a sand-filled calculating tray. It's the shape of something missing.

When we learn to count, the number words become a sort of sing-song. We memorise the sequence of counting numbers both in the form of sounds and in the form of symbols. I remember at a very early age practising this kind of recitation counting, going faster and faster, proud that I could count to 20 and beyond in a single breath. Other word sequences that we memorise are also disguised forms of counting: the days of the week, the months of the year, even the alphabet.[5] Many nursery and folk rhymes are counting rhymes: 'One, two, buckle my shoe', 'One for sorrow, two for joy'. The link between counting and recitation runs deep.

Feynman can count and read; Tukey can count and talk

The great physicist Richard Feynman interested himself in subjects way beyond his speciality: from lock picking to bongo drumming and throat singing, Feynman took delight in them all. In his book *What Do You Care What Other People Think?* as told to Ralph Leighton, he described experiments he did with counting: experiments to try to understand better how his own mind worked.

Feynman tried to train himself to count reliably to 60, in a consistent time. It didn't have to take a minute, he was only seeking consistency, and he found that he could keep to a reliable time of 48 seconds. Having established reliability, he then tried to disrupt his own mental process of counting. While doing his 'counting to 60', he raised his heart rate through exercise, he made himself hot, but nothing he tried disrupted the pace of his counting.

Until he tried counting his laundry, and his socks in particular, while keeping his counting to 60 going. He found that so long as he could visually recognise the number of items (using subitising and pattern recognition), he could determine the number without disrupting his internal count to 60:

I could write down '3' in front of "pants" or '4' in front of 'shirts', but I couldn't count my socks…

It's unsurprising really—the counting circuitry of his brain was already occupied. He found that by playing visual tricks, grouping socks by fours, he could do the

---

[5] When we assign letters of the alphabet to, say, parts of a document (Appendix A, Appendix B, …), we are in effect counting with letters.

arithmetic and calculate how many socks there were, using the subgrouping approach, but he couldn't do actual serial counting.

He tried counting lines of a book using similar visual tricks, and was successful. He could even read the text and still keep his counting to the consistent time of 48 seconds. Only when he tried to read out loud was the process disrupted.

When he reported these experiments to a friend, statistician John Tukey, one morning over breakfast, Tukey was disbelieving. Tukey maintained that reading while counting would be impossible, but by contrast, he felt that counting while talking out loud would be simple. Feynman, in turn, expressed his astonishment. And so each demonstrated to the other that their claims were indeed correct, for themselves.

Feynman:

We talked about it a while, and we discovered something. It turned out that Tukey was counting in a different way: He was visualizing a tape with numbers on it going by. He would say, 'Mary had a little lamb', and he would watch it! Well, now it was clear: He's 'looking' at his tape going by, so he can't read, and I'm 'talking' to myself when I'm counting, so I can't speak!

Naturally, neither could read aloud while counting mentally: Tukey was blocked by having to read, Feynman by having to talk.

Very small numbers, those we can subitise, may have a direct mental representation, but beyond those small numbers, our internal representation of numbers reflects the ways in which we interact mentally with them. Some people are like Tukey, and have mainly visual mental representations (like number lines and shapes), while others are like Feynman, and have mostly auditory representations (the sound of counting). Still others may employ other senses, for example tactile mental images.

I can relate to the Feynman/Tukey distinction: I am an amateur saxophone player and one exercise recommended for strengthening the embouchure is to alternately make the mouth into shapes for saying 'eee' and 'ooo': wide and pulled back, then round and pushed forward. The recommended exercise is to do this 50 times. When I tried this, I found a problem. I couldn't mentally count while moving my mouth in this way, because the shape change in my mouth from 'ooo' to 'eee' is very like the way my mouth moves when I say 'one'. The embouchure exercise kept resetting my counter to one!

So I must be, like Feynman, a talking counter.

Singing is counting

Music is the pleasure the human mind experiences from counting without being aware that it is counting. Gottfried Leibniz

We smile at the idea of the fellow entrusted with the triangle in a musical performance, who must wait for many bars until his moment comes. Before then, he must simply 'rest', and then strike his single note at the precise instant. How does he manage to keep count?

I play in an amateur jazz group. Many of the jazz standards we play have a formulaic 32-bar structure. Such a piece consists of repetitions of a structure called a chorus. It's 32 bars of 4 beats each. A single performance of the piece will typically consist of playing the chorus multiple times, each time following the 32-bar structure.

In performance, the group plays the melody (the 'head') right through, and then the rhythm section keeps right on going, chorus after chorus, while the soloists take it in turn to do their improvisations, each taking 32 bars. When all the solos are done, the group again plays the 32-bar head. In our group, none of this is written down: the large-scale structure is decided on the fly. Typically, all each player has by way of written music is a single sheet of 32 bars containing the chord progression and the notes of the bare melody, and sometimes not even the notes.

It's important to know where you are in the tune overall. You need to start your solo at the right time, and you all need to come in for the ensemble playing of the head for the last time round. This means that, in theory, everyone should be counting beats and bars.

In fact, no one is counting $32 \times 4 = 128$ beats to get to the end of each chorus. Most of us are not consciously counting at all, and that's because the piece has **structure**.

A jazz musician who has played 32-bar songs thousands of times develops a feel for counting the 128 beats in each repetition of the chorus. She will locate herself in the piece by understanding the geography of those 32 bars. Typically, a jazz piece has a recognised **form**. For example, one common form is 'AABA', which implies four 8-bar sections. The 'A' sections are substantially the same, but the 'B' section, 'the middle eight', differs harmonically and so sets up tension, to be resolved by the final 'A' section. Any moderately competent jazz musician will be aware of the harmonies being played by the rhythm section, and will 'know' where they are in the chorus at any time.

At a finer timescale, every player will have a mental metronome counting off beats in groups of four. For some, this will show itself in physical motion such as tapping of foot or nodding of head, or indeed strumming of guitar. For others, it will be just be happening in their head.

With some experience, we start to internalise these counting activities. Amazingly, most of the time, we are all able to hit beat one of the final chorus.

So counting has a close connection with rhythm and with measurement of time, and being able to keep count in this real-time way relies on being able to recognise larger-scale structures (in the case of music, not just beats, but bars, sections, and choruses).

Keeping tallies

This strategy of grouping items into larger-scale structures, bundling 4 beats to a bar, and so on, introduces the next wrinkle in how we go about counting to bigger numbers. It's certainly possible to count systematically to moderately big numbers (the hundreds, say) in your head, but as the number increases, the process becomes error-prone, and vulnerable to distraction. In practice, it makes sense to set a limit to how far you count without stopping to make a mark or keep some sort of record. Our pockets run out of pebbles.

In the same way that subitising stops being effective at some point, so there is a point at which serial counting in your head stops being effective. Mathematically, counting can go on forever—the counting numbers are an infinite set. But the human activity of counting, as an actual thing to do, runs out quite soon. Children love to demonstrate their ability to count: to 20 perhaps, or to 100, but not much further. Systematic serial counting has its limits.

If the counting process is liable to be disrupted, or spread over a long period of time, we need a means of record-keeping to keep on track. The cartoon prisoner counts the days of his sentence by tallying. Every fifth day, he draws a line through the marks he made on the previous four, the idea being that he can count the bundles of five, and then add to this the count for any remaining unbundled days.

When bank tellers count money, they typically count banknotes into groups of 20, making bundles which they then set aside and count at the end of the process. Election tally-counters, too, make bundles of votes for each candidate, the bundles then being assembled into bigger and bigger units to make packs of 500 (in the UK) which are then counted to arrive at a final result in the thousands. This tallying process means that any interruption only disrupts a small part of the process, and facilitates restarting and checking.

Tallying, working with bundles and remainders, takes us a long way and equips us for many aspects of life. With this skill, we can balance up a cash register at the end of the day, we can manage a stock-taking operation, we can take receipt of bulk deliveries.

We don't trust our own brains to remain accurate in our counting once we get to big numbers—nor should we. It is by sticking to a process, coupled to an organising scheme, that we are able to make the jump from 1, 2, 3 to being able to count into the thousands. While we don't trust our monkey brains, we do trust the consistency of the system. When we bundle 20 votes together, and then another 20, and have 7 remaining, we trust in the laws of arithmetic to assure us that the number we arrived at by adding numbers is the same number we would have counted to had we actually counted all the way through to 47.

Approximate counting

Counting is, in theory, an exact process. After all, the fundamental act of counting is '…and one more makes…', and if that one more didn't count, it wouldn't be counting, would it?

But, at some point, this simplistic approach gets hard to do. Somewhere beyond a thousand, for everyday purposes, we no longer feel a need for accuracy down to the last digit. We become content with an approach that just counts the bundles, and forgets the remainders.[6]

How big is Yankee Stadium?

If I'm told that the capacity of Yankee Stadium is 49,638 (as Wikipedia tells me), I might well consider that the odd '38' or even the odd '362' (the shortfall from 50,000) are unimportant in how I think about that number. I'd be happy to call that '50 thousand, more or less'.

And, just like that, we've shifted to a whole new mental strategy of coping with big numbers: we've approximated the number 'to the nearest thousand'. In doing this, we've effectively split the number into two parts: the digits we

---

[6] Naturally, if you're counting votes or if there is a specific need for precision, then you try to be as precise as possible. But for appreciating scale and doing 'more or less' calculations, we can be more relaxed.

think are significant (50), called the 'significand' (or 'significant figures'—and in this case the zero in '50' is significant) and the multiple (thousand) that tells us the 'scale' we are working at. These two parts give us a composite under-standing of the number. The scale tells us the order of magnitude we are dealing with—'what page we're on'—and the significand gives us a more accurate location of the number on that imagined page.

Should I believe Wikipedia? Can I accept that 50,000 is a plausible number as the capacity of a baseball stadium, even if I know nothing about Yankee Stadium? Just as a **cross-checking** exercise, and a workout for my numerical chops, can I form a view of whether or not that's a big number, when talking about baseball stadiums?

The question arises: Can I directly visualise a 49,638-seater stadium? The answer is no, not just like that, without thinking it through. After all, 49,638 is a big number.

Can I visualise 1000 seats? Yes, I can. I imagine a block of 25 rows of 40 seats each, a bit like a large movie theatre or a medium-sized concert auditorium.

Can I imagine 50 of those 1000-seat blocks? Yes. With some imagination I can.

I have no idea of what Yankee Stadium looks like or how it is laid out, but I can now start to construct a visualisation. For a start, the seats won't all be on the same level. I'll imagine that there are 3 tiers of seats, and that each tier con-tains 18 'blocks of a thousand'. That comes to 54 blocks, but it seems reasonable to imagine that there are 4 blocks lost on the ground level, for entrances and facilities, perhaps.

What do these tiers of 18 blocks look like? Of course, a baseball field is a dia-mond—so each tier might be arranged as a kite shape, with four sides, two of them 5 blocks long, two of them 4 blocks long, the seats being relatively more concentrated at the more desirable end. Maybe the blocks are separated by flights of wide access stairs, with vendors walking up and down selling lemon-ade, peanuts, and hot dogs. I can almost smell those hot dogs…yes, I can imagine this stadium!

Now I have a vivid picture in my mind, and that picture (3 tiers, four sides of 4–5 blocks each, each block of 1000 seats) seems reasonable. The visualisation suggests a big stadium, but not huge. Am I right?

If I now try to compare this stadium's capacity with others, I'm able to think in terms of thousands of seats (as long as all the other stadiums are measured in thousands, they'll all be on the same page, so they'll be directly comparable). That means I can easily compare the 50 thousands of Yankee Stadium with the 56 thousands of Dodger Stadium, which is the biggest baseball stadium in the US.

Yankee Stadium comes fourth on the list of stadiums in the US. Fenway Park in Boston (which I *have* visited) is 28th on the list and has a capacity of 'more or less' 38,000.[7] The biggest (non-baseball) stadiums in the world have capacities somewhat over 100,000. So the conclusion that we reached, that 'when it comes to sports stadiums, 50,000 is big, but not huge', seems to be about right.[8]

Think through what we've done in this mental exercise. We've employed two powerful tools:

We have reduced the big number of '49,638 seats' to the very manageable '50 blocks of 1000'. The cost of performing this trick is that we have distanced ourselves from a raw appreciation of the things we are actually counting, the seats, and have replaced the hard image of a seat with the fluffy image of a notional block of a thousand seats. This seems a small cost, but we should always remember that we are now dealing with multiples of a thousand. At the level of thousands, that's no great burden, but you should not dismiss this 'distancing' effect. After all, our idea of exactly what a thousand seats look like might not be so accurate. And down the line, when we use this trick, as we will, for even bigger numbers, the multiples will become millions and billions and trillions, and those bigger multiples have a way of getting more and more vague in our minds, and more and more difficult to keep a grasp on. But, for all those reservations, the power of this approach makes it one of the five techniques central to this book. I call it **divide and conquer**, and we'll come back to this strategy many times.

We have also explored the power of another of the five techniques: **visualisation**. By using simple arithmetic, we have built a model of Yankee Stadium in our heads and that has done the work for us of confirming that the suggested number of 50,000 seats is indeed credible—more or less. Whether the number quoted on Wikipedia is precisely correct or not, we have satisfied ourselves that the number is at least plausible. And that's at the heart of understanding big numbers.

---

[7] Now having a reference visualisation in the form of Yankee Stadium, I can visualise Fenway Park as being '12 fewer blocks of a thousand each' than Yankee Stadium. And that is one block fewer on each side, on each tier.

[8] And having worked through this example, I now have in my head a couple of '**landmark numbers**' for stadium size. If someone tells me of a stadium that has a capacity of 150,000, I can gasp in amazement (or challenge them).

When is 'more or less' good enough?

It's not the voting that's democracy; it's the counting.   **Tom Stoppard**

In an election, it is quite common to have to recount the votes. When the result is within a margin of error (in the UK Parliamentary Elections, a margin of 50 votes would usually be close enough for a call for a recount), the votes will be recounted until a decision that is beyond legal challenge is reached.

The tacit assumption here is that in all elections the count will be imperfect. There will be always be some mistakes, some imprecision in the count, but so long as the margin of victory far exceeds the likely margin of error, we're happy to disregard the fact that the numbers quoted may not be exactly correct.

When a census is taken (which after all is nothing more than a complicated counting exercise), the statisticians allow for a proportion of non-responders and other myriad sources of imprecision, and settle on a best estimate.

The website of the US Census office has a population clock. It purports to show the total number of people in the USA, and the population count ticks over like an odometer counting the miles in a car. Naturally, the US Census does not operate in real time: they work from a series of ten-yearly censuses, supplemented by annual 'measurements'; from these, they derive a set of monthly estimates and projections, and they set the rate at which their clock ticks over, based on an average rate for the current month.

The population clock then, though purporting to be counting people, is simply measuring time in units of people. So, while Feynman uses counting as an estimate of time, and the jazzers measure time by counting beats and bars, the US Census uses time as a proxy for showing population change.

For most purposes, when counting numbers get very large indeed, we're content to work with approximations. For everyday numeracy, we don't need accuracy to the last digit. Of course, we'd like the first few digits of the number to be correct, and, most important, we want the order of magnitude to be correct. Counting populations for census purposes may be hard, but some populations are even more slippery.

How many fish in the sea?

The promotional poster for the movie *Finding Nemo* claimed that there were 3.7 trillion fish in the sea. Where does that number come from? Not by counting fish one by one, that's for sure. A number like this can only ever be an estimate, but what goes into that process of estimation?

The answer is modelling and sampling. Scientists construct a 'model' that divides the seas and oceans of the world into multiple regions, and includes details of which species of fish are expected to be found in each region. Then they sample as many regions and volumes of the sea as possible, in many different ways.

Samples are taken by methods such as looking at records of fish landed by commercial fishers, as well as dedicated surveys by research ships, to come up with estimates for each part of the model. A calculation is then done, totalling all the partial elements and allowing for unknowns, to come up with a number that is a 'best estimate'.

It's good to do a **cross-comparison**, and, aspiring to be numerate citizens, we can look for other ways in which such an estimate can be made. For example, in 2009, a researcher at the University of British Columbia did a study that looked at ocean plant production and how it progresses through the food chain. She came up with a total amount of fish biomass of between 0.8 and 2 billion tons.

So we might take the midpoint of that range (1.4 billion tons), and if we were to judge that each fish weighs on average half a kilogram, that gives us 2.8 trillion fish. This is close enough to allow me to conclude that *Finding Nemo*'s 3.7 trillion estimate is at least of the right order of magnitude. More or less.

How many stars in the sky?

Counting the stars individually is impossible, naturally, but, as with the fish in the sea, an approach using modelling, sampling, and calculation allows for a best estimate that gives us the approximate order of magnitude.

For example, very roughly speaking, astronomers estimate that the average number of stars in a typical galaxy might be between 100 billion and 200 billion stars. Furthermore, it is estimated that there are around 2 trillion galaxies in the observable universe.[9] That means that there must be between 200 and 400 sextillion stars.

Alarm bells: 'sextillion'? Million, billion, trillion—these are words I don't need to think too much about, but sextillion is something I need to work out in my head each time I come across it. A sextillion is a thousand billion billion, 10^{21}, and it is a very big number indeed. It's pretty much at the limit of where it is helpful to call numbers by the '-illion' names. So, instead we turn to scientific notation, and say that the number of stars in the observable universe is probably between 2×10^{23} and 4×10^{23}.

---

[9] For the first draft of this chapter, I used a figure of 200 million galaxies—the commonly accepted best estimate at the time. However, in October 2016, results from years of study using the Hubble Space Telescope were released that increased the best estimate by a factor of 10.

> **Landmark numbers**
> - Low-ball estimate of stars in a galaxy: 100 billion
> - Estimate of galaxies in the universe: 2 trillion
> - Low-ball estimate of stars in the observable universe: 2×10^{23}

So, what counts?

It turns out that counting is not so simple after all. Counting 1, 2, and 3 seems to be innate, but beyond that we either settle for approximate impressions of numbers (numerosity) or we rely on mental strategies of matching against memorised sequences, either auditory or visual. But even these systematic counting techniques run out quite soon and we have to introduce record-keeping. For really big numbers, there is no actual counting at all: we make models, we sample elements of those models, and we compute the count. And for the biggest numbers of all, we count ourselves lucky even to get the order of magnitude right.

How big is a billion?

Peppered through this book you'll find pages like the ones that follow. I call them 'number ladders'. I'll take a starting number or measurement (in this case 1000), and give an example (or two) of something in the real world that 'more or less' matches that number. Then we increase the number in steps so that after every 3 steps we reach a number that is 10 times larger. And repeat...

1000	Number of patents held by Thomas Edison = 1093
2000	Number of paintings that Picasso painted =1885 Population of Norfolk Island[10] = 2200
5000	Number of container ships in the world = 4970 Population of Montserrat[11] = 5220
10,000	Population of Cook Islands[12] = 10,100
20,000	Population of Palau[13] = 21,200
50,000	Population of Faroe Islands[14] = 49,700

---

[10] A tiny Australian island in the South Pacific Ocean.
[11] A mountainous Caribbean island.
[12] A South Pacific archipelago nation of 15 islands.
[13] An archipelago of more than 500 islands in the Western Pacific.
[14] Okay, more islands... I'll stop now

100,000   Population of Jersey = 95,700
               Number of seats in Melbourne Cricket Ground = 100,000

200,000   Population of Guam = 187,000

500,000   Population of Cape Verde = 515,000

1 million   Population of Cyprus = 1.15 million

A million and up...

1 million   Population of Cyprus = 1.15 million

2 million   Population of Slovenia = 2.05 million

5 million   Population of Norway = 5.02 million

10 million   Population of Hungary = 9.92 million

20 million   Population of Romania = 21.3 million

50 million   Population of Tanzania = 50.7 million

100 million   Population of the Philippines = 99.8 million

200 million   Population of Brazil = 202 million

500 million   Estimated world population of dogs = 525 million

1 billion   Number of cars in the world = 1.2 billion

Just look at how much countries differ in terms of population size: for example, just from the list above, Brazil (around 200 million) is about 10 times the size of Romania (around 20 million), and 100 times the size of Slovenia (2 million). Wouldn't those make good **landmark numbers**?

A billion and up...

1 billion   Estimated world population of domestic cats = 600 million

2 billion   Number of active users of Facebook (June 2017) = 2 billion

5 billion   Number of base pairs in the human genome = 3.2 billion

10 billion   Population of the world = 7.6 billion

20 billion   Number of chickens in the world = 19 billion

50 billion   Number of neurons in the human brain = 86 billion

100 billion   Estimated number of humans who have ever lived =
                 106 billion

200 billion	Number of stars in our galaxy = 200 billion
500 billion	Number of neurons in an African elephant's brain = 257 billion
1 trillion	Number of stars in the Andromeda galaxy = 1 trillion
2 trillion	Number of trees in the world = 3 trillion
5 trillion	Number of fish in the sea = 3.7 trillion
10 trillion	Number of bits in a 1-terabyte capacity hard drive = 8.8 trillion
20 trillion	Number of human cells in the human body = 30 trillion
50 trillion	Number of bacteria in the human body = 39 trillion
1 quadrillion	Number of synapses in the human brain = 1 quadrillion

More than that and we're getting to astronomical numbers—and there's a separate chapter for those!

More spurious coincidences in counting up

Population of **Japan** (126 million) is
 2 × population of the **United Kingdom** (63.6 million)

Number of rivets in the **Eiffel Tower** (2½ million rivets) is
 2 × population of **Swaziland** (1¼ million)

Number of seats in **Wembley Stadium** (90,000 seats) is
 25 × number of seats in **Hammersmith Apollo** (3630 seats)

Active personnel in the **Indian armed forces** (1.325 million) is
 20 × seats in **Eden Gardens Cricket Ground** (India) (66,000)

Automatic Teller Machines (**ATMs**) in the world (3 million) is
 2 × estimated world population of **moose/elk** (1.5 million)

High estimate of world population of **lions** (47,400) is
 the same as the high estimate of **spotted hyenas** (47,000)

Numbers in the World

How Numeracy Connects to Everyday Life

Everything around you is mathematics. Everything around you is numbers.

Shakuntala Devi

Which of these weighs the least?
☐ medium-sized pineapple
☐ A typical pair of men's dress shoes
☐ A cup of coffee (cup included)
☐ A bottle of champagne

The age-old question: How strong is the beer?

In the early 1890s, in the ruined Egyptian city of Thebes, the Russian Egyptologist Vladimir Golenishchev bought a papyrus scroll. It was 5.5 metres long and 7.6 centimetres at its widest, and it contained 25 instructive problems in arithmetic. The papyrus has been dated to around 1850 BCE, and is now known as the *Moscow Mathematical Papyrus*.

The scroll contains a range of exercises for students, and covers a variety of topics. Two of the problems have to do with calculation of proportions for parts of a ship: a rudder and a mast. Another of the problems deals with calculating the volume of timber in logs; yet another has to do with the output of a worker making sandals. Other problems are in geometry and include the calculation of the volume of a truncated pyramid.

But 10 of the 25 problems concern themselves with calculations to do with baking and brewing. How much grain must be used for the making of specific quantities of bread and beer? And, most important, what calculations are needed to predict and thereby control the strength of the beer that is brewed?

Perhaps some things will always be priorities. A functioning, complex society will always need numbers. And Ancient Egypt needed its numerate scribes to manage those numbers, not just for the building of pyramids, but right down to everyday matters such as the brewing of beer.

What is numeracy?

It's not mathematics

The website run by Teach in Scotland had, at the time of writing, this apt definition of numeracy:

Numeracy is the everyday knowledge and understanding of number and reasoning skills required to access and interpret the world around us. We are numerate if we have developed confidence and competence in using number which will allow us to solve problems, analyse information and make informed decisions based on calculations.

Beware: this is not a description of mathematics. Mathematicians are certainly very interested in numbers and the concept of number, and Number Theory is a specific and important branch of mathematics, but don't be misled by the name. Number Theory is a very different thing from numeracy.

The essence of mathematics is the abstract thinking that lets you reason in a rigorous way about mathematical 'objects'. Mathematical objects are themselves abstractions (and some of them are numbers, but there are many other types of mathematical object).

Numeracy, by contrast, is a practical capability. It's to do with how the abstraction of numbers connects to the physical world of objects and the practical world of social interaction. The definition provided by Teach in Scotland brings out some of these features. It stresses:

- 'Everyday knowledge and understanding'
  Numeracy is for everyone, and for everyday use.
- 'The world around us'
  Numeracy is connected to the day-to-day world.
- 'Confidence and competence'
  Numeracy should be a familiar tool, a sharpened blade, close at hand and ready for use.

- 'Make informed decisions'
  Every citizen is a decision maker, not least in their capacity as a voter. Decision-making needs understanding. Understanding needs knowledge. Knowledge needs numeracy.

It's a cliché of our times that, while admitting to illiteracy would be an embarrassment, innumeracy ('I'm no good with numbers') is almost seen as a badge of pride. I'll assume that, if you're reading this book, I needn't spell out how absurd this is. Rather than ranting against innumeracy, this book celebrates numeracy and the delight that comes from developing and using a fluency with numbers in day-to-day living.

It's not accounting

Numeracy is not mathematics, but neither is it about being good at arithmetic (though that might be a side effect). We don't all have to be bookkeepers to be numerate, or to be able to add up long lists of numbers without error.

The skills of a numerate citizen are not so much to do with being able to account for all the pennies as with knowing whether the pounds amount to a big number, or a small number, in whatever context is relevant.

Folk numeracy versus scientific numeracy

Numeracy arises naturally from life. Rushing to catch the train, we're comparing numbers and estimating speeds. Planning a meal, we're judging quantities. Watching sports, we're absorbing and evaluating stats. Even our hunter-gatherer ancestors would have been tallying the catch for the day, making sure it would suffice to feed the group; judging if they could make it back to the cave by nightfall; counting the days before the spring thaw would come.

Numbers are woven into our cultures. The Bible itself is full of numbers and measurements. Indeed the fourth book of the Old Testament is called *Numbers*, referring to the two censuses conducted in that account of the wanderings of the Israelites. Our language bears witness to the numbers and measurements in our lives and those of our ancestors: a fortnight is 'fourteen nights', a 'furlong' is a 'furrow's length', the up-and-coming generation are the 'millennials', we achieve 'milestones'. Even children's rhymes have their hidden numeracy: it's no coincidence that 'Jack' and 'Gill' are both measures of liquid capacity.

The numeracy needed to get on with life is natural, and is necessary for daily activities. When the shepherd counts her sheep, when the miller bags his flour, when the innkeeper pulls a pint and counts the pennies, there is no sense of this level of numeracy being a specialist skill. There's no 'I'm not a maths person', no fear of numbers. On familiar ground, at human scale, numeracy is natural. This is folk numeracy.

But there comes a point where the community starts to need specialist skills. The village becomes a town, and the authorities demand taxes. Now the tax-collector must assess and write down how much is due from hundreds of households, and so the numbers start becoming bigger. As the sophistication of the economy develops, there is a need to deal with the much bigger numbers of state economies and populations. Those who study mathematics and the sciences are admitted into a kind of priesthood of higher numeracy, and they learn to deal in a more universal way with numbers that go far beyond the needs of day-to-day commerce and coexistence.

The link with everyday life is broken: when the tax collector collects taxes for a city, there is no real connection between the totals collected and the impacts on individual households. People become ciphers. The amounts to be gathered and the plans for spending become a budget, and the numbers in the budget are big numbers.

Even those who have not entered the financial or the mathematical worlds are affected by those big numbers, those nation-scale, those astronomical numbers. Democracy asks us all to make decisions that demand some understanding of the size of national budgets, the impacts of human activity on the natural world, or the consequences of political decisions on trade and wealth. Very few of us have an adequate understanding of the numbers required to make truly informed decisions.

We're innumerate, but literate

Even stranger things have happened; and perhaps the strangest of all is the marvel that mathematics should be possible to a race akin to the apes.

Eric T. Bell *The Development of Mathematics*

As a species, we have very little inherent numerical ability. What ability we have to directly perceive and 'sense' numbers is essentially restricted to two skills: that of 'subitising', the direct and instant perception of numbers without counting or pattern recognition (and that ability ends somewhere around four items[15]), and a number approximation sense (a sense of numerosity) that allows for an imprecise impression of somewhat larger numbers.

And yet, jumped-up apes that we are, we have, collectively, managed to reach a level of numerical sophistication sufficient to send a space probe five billion kilometres to Pluto and have it arrive at the right place and at the right time. How on Earth (or away from it) have we magicked our rudimentary sense of number into this formidable arithmetic capability?

The answer is that we've done this by using a range of other intellectual capabilities of our minds (primarily skills associated with language, organisation and philosophy). We've evolved mental strategies and tricks that enlist these other capabilities into the service of numeracy. Our language itself makes clear the entanglement between numeracy and literacy: the person counting money in the bank is the **teller**; the story-teller **recounts** her tales.

Counting is singing. Singing depends on memory and repetition and rhythm— exactly the skills that are needed for counting. The simplest counting relies on memorising a sequence of numbers, and then simply matching, one-by-one, the things being counted, against the memorised sequence. When the sequence runs out, we repeat it, perhaps with variation ('one, two' becomes 'twenty-one, twenty-two'), as many times as is necessary.

We store up memories deliberately. Just as the bard learns the epic poems, so we coach ourselves in the times tables to provide the internal resource that we can then later use to fuel the algorithms of mental arithmetic.

The stories we recount often work in formulaic ways, for example setting up explicit contrasts: the good son and the prodigal; Gandalf the Grey and Saruman the White. The Sun rules the day, the Moon rules the night, and we learn about symmetries and balance and pattern. We assess and compare, and we turn these skills to dealing with numbers too.

The glamorous sister of memory is imagination. Imagination lets us make new songs and new stories, but it also lets us see the future. We can visualise the not-yet-sown field, and work out how much seed it will take to sow, and we can plan for the harvest we expect it to yield.

We humans can cooperate, and organise ourselves. We can make plans, detailing the steps in a complex process, and then follow those steps. In this way we learn to tackle jobs in a systematic way, knowing which tasks must follow which. In the same way, we devise complex chains of thinking about numbers,

---

[15] By contrast, a computer, even a pocket calculator, has numerical abilities inherent in its construction—literally hard-wired.

using the same organisational skills to keep track of where we are in the computation.

None of these skills is inherently numeric, but we turn them to use with numbers. We are cunning monkeys.

But we're not just cunning monkeys. We can be wise old apes as well. We know how to call forth the abstract spirits of number and shape and structure from the world around us. We argue and we reason, and from our arguments and our reasoning we abstract the skills of logic, and formalise the processes of thinking.

Driven by our deep curiosity, we break things apart, physically and conceptually, and we make mental models and abstract structures: wholes can be broken down into parts, names can be given to the parts, names can even be given to the **relationships** between parts.

We make journeys and we tell stories, and from this journeying and this story-telling we learn concepts of sequence and consequence, and from these we abstract the notions of chains of reasoning. Equipped with abstraction and structure and logic, we have acquired the skills to become mathematical.

But this is not a book about mathematics. This book is about clearing the bewildering fog that big numbers seem to bring, that clouds our view of the world. How do we clear this fog? We're not all going to have our brains re-wired to become savants, nor will we take some wonder drug to open new pathways in our brain.

In fact the elements we need to develop our numerical thinking are already in place: memory, sequence, visualisation, logic, comparison, and contrast. The cultural skills that make us human can be used again in the service of greater numeracy.

That's why you'll find this book sprinkled with numbers that I call landmarks: memorable and useful. You'll also find number ladders, sequences of ever-increasing numbers, making a strange kind of numerical poetry. There are stories telling why we use the words we do when talking about numbers. You'll find visualisation examples to help you build castles in the air. You'll find oddball comparisons and contrasts, to dramatise differences.

So, although we may be inherently innumerate, with less native calculating ability than an abacus, nonetheless we're very clever at turning our cultural strengths to numerical purposes. When we add to this cleverness the ingenious mechanical aids we have invented, the technologies of writing, of calculating machines, and computers, then we succeed in raising our number-processing capabilities to the point where we can build the machines that literally fly through the heavens and take us to new worlds.

Words about numbers

Why do words matter?

Why bother about words in a book about numbers? Because this is also about how we think about numbers, how we sense numbers, and how we feel about numbers.

Words are not just arbitrary labels. Words carry within them the traces of their origins and evolution, they retain connections to their past. This is also true of number words, words with very deep roots. Consider the first few counting numbers, and the weight of implication and association they carry:

'**One**' encapsulates the most fundamental question there is, about the existence of the universe: 'Why is there something rather than nothing?' 'One' is precisely the difference between something and nothing. 'One' is the number of self. I am one body, I have one 'point of view' in my head from which I perceive the world, I walk one path through my life, I see one world, one horizon. 'One' is the start, 'one' is unique.

'**Two**' is the first multiple; it is the first instance of the concept of 'more than one'. 'Two' is the first number that truly conveys the idea of 'number': it is the first plural. 'Two' is the first number that can characterise relationships between things, that can signify a bond, a coming together or, indeed, a splitting apart. 'Two' is the first number that can talk about symmetry (two hands, two feet). 'Two' is the first number that can describe opposites—East and West. It is the first number that can characterise difference ('this one' versus 'that one'). All other counting numbers must tread in two's footsteps.

'**Three**' is rich in mystical and magical connotations. 'Three' is the number of creation, the possibility of two things combining to produce one—Mum, Dad, and child; thesis, antithesis, and synthesis; numerator, denominator, and quotient. The three bears: big, small, and in-between. 'Three' is Trinity.[16] 'Three' is the number of spatial dimensions we perceive. Three legs on a table are the minimum that will give stability; three straight lines are needed to draw a geometrical shape that encloses an area. Rhetoric

---

[16] At Trinity College, Cambridge, the punts taken out on the River Cam all have names that associate with the number 3: 'Hat Trick', 'Diabolus in Musica', 'Musketeer', 'Mad George', 'Harry Lime', 'Menage', etc.

uses the rule of three because three is the smallest number that can establish a rhythm. 'Three' is the first number that can express exclusion.

'Four' has strong associations with sturdiness and order. The cycle of the year—growth, maturity, decline, death. Four legs on many animals, four compass directions. 'Four' implies squares and rectangles, and buildings and fields, and square dealing and tidiness.

'Five' fingers on my hand, five toes on my foot. Five days in the working week.

And so it goes. The first few counting numbers mean much more than their bare numerical significance: they are freighted with meaning.[17] These additional meanings are hooks on which we can hang associations. A family with four members? We can see them seated around a square table. Three things on the shopping list? In my mind, I make a triangle of them.

Words embracing numbers

Numbers sneak into our language everywhere. If you 'atone' for a crime, you make yourself 'at one' with it. The Indo-European root dwo- (meaning two) is responsible for multitudes of two-related words, including duels and duals, two and twain, dilemmas and dichotomies, and even dubious duplicity. 'Three' gives us trivia, trident and tripod, tricycles and trigonometry. Visiting the University of St Andrews some years ago, I spotted a sign reading 'Hebdomadar's Block': the

---

[17] In fact, the first few numbers are so special that we break all sorts of naming rules for them. We talk of 'first, second, third', not 'oneth, twoth, threeth'; and 'eleven, twelve, thirteen', not 'oneteen, twoteen'; 'twenty, thirty', not 'twoty, threety'. A 'one' in cards is an 'ace' (and a two is sometimes a 'deuce').

word derives from the Greek root for 'seven' and is title given to the official in charge of discipline for the week.

If we're lucky, we have twenty/twenty vision. If we're unlucky we might be in a Catch-22 situation. A round of golf might end up at the nineteenth hole, after which we might end up three sheets to the wind.

Etymology of unit names

The Greeks had 'podes', the Romans had 'pedes', and we have 'feet' as a basic unit of distance. These measuring feet are of course in the first place derived from actual human feet. While a foot-long foot would be a rather large foot, there is still a degree of naturalness and comfort with the 'foot' as a unit of measurement. It is human scale, and human-scale measurement units like the foot are naturally more intuitively understood. These human-scale measurements have been termed 'anthropic units' and we'll come across many more of them.

'Full fathom five thy father lies', sings Ariel to Ferdinand in Shakespeare's *The Tempest*, but few of us now know how deep a fathom is. But if we understand that a 'fathom' originally came from a word meaning 'outstretched arms', it becomes much easier to remember that a fathom is 6 feet (equally, 2 yards, the yard being more or less as long as nose to tip of outstretched arm).

I find this fascinating for its own sake, but even more so for how it knits together the language, ideas, and numbers of life. The connections make things memorable, and when you remember things, you can use them to make sense of the world.

Scientific notation

The standard scientific 'exponential' notation for large numbers is well adapted to the purposes for which it is used, namely scientific and engineering calculations. It is a universal system that can be turned to measurement tasks ranging from the unbelievably small to the truly astronomical. We'll not be able to reach the end of this journey and discuss really big numbers without using this notation sooner or later. Here's a reminder of how it works:

So, in scientific notation, the population of the world (7.6 billion) would be written as 7.6×10^9, the 10^9 signifying that a billion is the 9th power of 10, that is a 1 with 9 zeros. Likewise, 1.5×10^{-14} m (note the negative power) signifies a very small distance, in this case the diameter of a uranium atom. Written as a conventional decimal, it would have 13 zeros after the decimal point: 0.000000000000015.

Why not simply use that system in this book? Well, it doesn't really suit our needs, because it replaces the sense of number with an intellectual decoding process. It's a system that is quite far removed from everyday life: it's the language of the priesthood. We'll need it when we reach the truly enormous numbers describing the size of the universe, where there really is no alternative, but for practical numeracy I'd prefer to use number labels with a bit of character: number names, not just numbers describing numbers.

So, for our purposes, we'll write the length of the equator as '40,000 kilometres'. By doing this, I hope you'll be able to use some sort of gut feel for what a thousand kilometres might be is like (it's a little more than you could comfortably drive in a day), and then I can simply ask you to visualise 40 of those 'standard distances', those long-day drives. It's another way we can use the **divide and conquer** strategy, with a little **visualisation** thrown in.

The powers of a thousand

Where do the big numbers start?

What defines a big number? Where does our comfort zone for numbers end and where do we lose our 'feel' for numbers?

Wading into the surf, I start off with my feet firmly on the sand, and can push forward through the water with some confidence. But going deeper, I find there comes a point where the currents start buffeting me, and I almost lose my balance. And still deeper, I reach a point where keeping one toe on the bottom just isn't working, where a new strategy (treading water, or swimming, or finding a flotation aid) is needed. I've reached my depth.

That's the sense I have when I wade into the sea of bigger and bigger numbers. There comes a point where straightforward visualisation and innate number sense starts to fail, my toe loses contact with the bottom, and new strategies are needed. If I want to venture further into the sea of numbers, I'll need to extend my range, and find ways to cope when I'm out of my depth.

But what is my number depth? Where do the big numbers start?

Firm ground

Subitising, that pure number sense of directly and instantly perceiving how many items there are in a group without counting or pattern recognition, ends around 4—if we rely on subitising, everything from 5 onwards is a big number. But at school we learn techniques that give most of us an intimacy with and a facility for dealing with numbers that extends at least into the hundreds, and maybe a bit beyond.

Those who choose a scientific path in their training are taught rigorous techniques for manipulating a notation for numbers that is essentially unlimited. Scientific notation and calculations based on this notation become practised and almost automatic.

But as the numbers get larger, we all—even the mathematicians and scientists among us—start to lose our innate feel for the numbers. The bigger the numbers become, the more we find ourselves out of our depth and the more we need to rely on algorithmic processes, applied by intellectual effort, for our understanding.

Take as an example a number like 2.5 trillion. Few people can have a ready-to-use mental image of this number: all of us must apply some intellectual processing to decode this and start to make sense of this number in whatever context it is presented.

Here are some options for digesting that number:

- Treat the 'trillion' as a unit, a black box, that if needed we could look inside, but we'd prefer not to. If we were told that the US budget had increased from \$2.2 trillion to \$2.5 trillion, we might be content with this option. In this case, the relative increase is where our interest lies.
- Use scientific standard exponential notation. Simple, says the scientist, this is 2.5×10^{12}, and I know how to work with numbers like that: I am very familiar with the algorithms needed to process those numbers.
- Apply per-capita style **ratios**. 2.5 trillion—well, that's easy to comprehend, says the economist. I know that there are 7+ billion people in the world, so if the cost of a global programme is \$2.5 trillion, then that works out to around 300 dollars for every person on earth.
- Make **landmark** comparisons. If you have the right knowledge and if the context is appropriate, you might say that 2.5 trillion is roughly 70% of the number of fish in the sea.

No-one can absorb such numbers without one or more steps of mental processing. It's simply a question of which mental strategies are most appropriate.

Where do I reach my numerical depth?

For me, the range of numbers I can confidently sense, or visualise, or intuit, or grok, runs out somewhere around 1000.

The highest ever innings total in a cricket test match was 952 runs scored by Sri Lanka in 1997 against India. We can visualise how that score accumulated: mostly in singles, twos, fours, and sixes, and we can imagine the score gradually totting up to that huge total. It's a big number, but I can grasp the reality of it. It's nearly at my limit, though. It's close to slipping from my grasp.

Naturally, when numbers get bigger, I have my repertoire of strategies, my flotation devices for coping with those bigger numbers. I can intellectually deal with, and mathematically manipulate, much bigger numbers, but beyond one thousand I am aware that I am relying on the algorithmic mental tricks I described above.

Similarly, if we look at small ratios and fractions, I'll draw the line at a thousandth ($^1/_{1000}$). Smaller than that I cannot meaningfully visualise: the number is below my mental resolution.

This is a personal observation about how I think about numbers, but I think I'm not unusual in this. There are clues that our culture has codified the way we talk and write about numbers in a way that suggests that this breakpoint of one thousand is widespread. Somewhere between the hundreds and the thousands, we find we need to shift gears, to start using different mental techniques to think about numbers.

Exposure to big numbers

At school, we learned our times tables, running to $12 \times 12 = 144$; our class sizes were measured in 10s, team sizes were 11 (cricket) or 15 (rugby). As a kid, I was triumphant about being able to count to 100, and once or twice made the effort to count to 1000. In this way, competence in primary-school mathematics took me to a comfortable familiarity with numbers up to around the range of a hundred, or a bit more.

Also at school, we learned the **definitions** of bigger numbers (millions, billions, etc.), and we flirted with the idea of infinity in the form of that wonderful non-number 'uncountable', but it would be wrong to say that I ever internalised the idea of a million. It was more that I understood and trusted that if I just kept going with the same counting process I had learned, I could eventually count to a million, just as I knew that walking for long enough would take me to the next town even though it was over the horizon.[18]

But in adult life we confront numbers that are very much bigger. Crowds at sports events might reach into the thousands or tens of thousands of people; we talk of numbers of refugees in terms of thousands and millions; national population statistics reach the millions and billions, and national budgets get to billions and even trillions of whatever currency applies.

And sometimes—often—these numbers are important: it's not good enough to plead ignorance or bafflement. Bad arguments get made and bad policies are implemented because people have no clear sense of what these numbers mean. 20,000 refugees seems like a big number, and in some contexts it is, but compared with a population of 60 million, it is just one for every three thousand people. And 1 in 3000 is a number that is just below my mental resolution: it's really too small for me to have a natural feeling for. I need some mental strategies to deal with it.

Visualising a thousand

If you're like me, somewhere between 'in the hundreds' and 'in the thousands' things start to get blurry and we shift from numbers we can see rather clearly to ones that start to fall into a 'big numbers' blur. Here are some illustrations of what a ratio of one in a thousand is like:

The sequence of black lines below shows a diminishing central gap, ranging from one part in ten at the top to one in a thousand at the bottom. You may struggle to even see the break: that's what a ratio of a thousand to one looks like.

---

[18] Of course, I learned many ways of understanding and manipulating the number 'a million': symbolically through arithmetic; seeing it as the cube of 100 and the square of 1000, but these are processes and not direct apprehension of the number.

| 1 in 10 |
| 1 in 20 |
| 1 in 50 |
| 1 in 100 |
| 1 in 200 |
| 1 in 500 |
| 1 in 1000 |

Some other examples:

- 1000 to 1 is the full length of Route 66 compared with the length of New York's Central Park. Today, the fastest drive from Chicago to LA would take around 4 days. By contrast, driving past Central Park (with no traffic) takes just 5 minutes.
- On a cricket pitch, the distance from wicket to wicket, set against a one penny coin. That's 1000 to 1.
- Visualise Kilimanjaro, the highest mountain in Africa: against that background, a giraffe. That's 1000 to 1.
- The length of an aircraft carrier, compared with a flea's jump. That's also 1000 to 1.

Powers of a thousand in language

It turns out that the English language agrees with me. Our numbering system is decimal, based on the number 10. When it comes to numerals, we add a new 'place' to the representation for each new power of ten. So, in terms of numerals, we are firmly in the world of base 10. We have words for the first few of these tenfold steps: 10 = 'ten', 100 = 'hundred', 1000 = 'thousand'.

But then? We have no single word for 10,000. We talk in terms of ten multiples of 1000. The language itself is signalling that we have now crossed into big-number territory, where the way to deal with the numbers is in terms of chunks: ten chunks of 1000 each. The difficulty of dealing with a number as big as 10,000 has led us to split the concept into two parts: a pseudo-unit of a thousand, and a significand (the significant figures) that forms a number within our comfort zone, 10.

Of course there **are** more number words beyond 'thousand'. We have 'million', 'billion', 'trillion', etc. Please note: these are all powers of 1000. This suggests that we are somewhat comfortable talking about a number like 728 thousand, but by the time we reach 21,352 thousand, it seems to be more natural to start talking in millions: we might restate this as 'about 21.4 million'. We renormalise the number to allow the significand to sit nicely in our comfort zone, and the word 'million' packages up the bigness of the number in a way that makes it easier to carry around.[19]

Even when we stick with digits, we recognise the importance of the thousand-fold breakpoint. A number such as 125,000,000 is conventionally written with commas separating it into, you guessed it, powers of 1000. The way we are accustomed to dealing with big numbers in everyday life clearly marks out the importance of 'one thousand' as central to the everyday talking about big numbers.

All roads lead to a thousand

Even the Romans have their part to play in this: their numbering system essentially runs out at M = 1000.

Beyond that, a bar over the number (such as $\bar{C}$) signifies that the number relates to a multiple of a thousand (in this case 100,000). They, too, recognised the change of gear that happens at 1000.

And it doesn't stop there. The metric system defines prefixes for:

10 ('deca-'), rarely used
10^2 ('hecto-'), rarely used
10^3 ('kilo-')—and then there is a gap until:
10^6 ('mega-')
10^9 ('giga-'), and so on.

Likewise for subdivisions:

$\frac{1}{10}$ ('deci-')
10^{-2} ('centi-')
10^{-3} ('milli-')—and then a gap until:
10^{-6} ('micro-')
10^{-9} ('nano-'), and so on.

So, even in the scientific world, the powers of a thousand are given special treatment.

---

[19] It's also interesting how the 'long' system of naming big numbers based on powers of a million—where a billion is a million million, and a trillion is a million billion—has effectively fallen into disuse (at least in the English-speaking world), despite being much more logical than the 'short' system we use. More logical because the 'counting' prefixes (bi-, tri-, quadri-, quinti-) actually make sense in that system, which they don't in the system we actually use. Go figure!

Precision

Engineers may need to work to extremely precise tolerances, but for your average builder or handyman, the precision needed for their work is limited. The bookshelf beside me is around 2 metres long and 1 metre high. It's a fine and well-made bookshelf, but if I measure it accurately, I come up with discrepancies of the order of 1 to 2 millimetres. That's a precision of 1 in a 1000.

On the web, you can find plans for home-built aircraft, proper planes intended to fly and carry passengers. I took a detailed look at one in particular: nowhere on those plans could I find an actual measurement that was more precise than three significant figures, and only a few numbers at all (they were the results of calculations) that went to four significant figures. I'd expect that the aircraft engine itself would have been assembled by specialists working to finer tolerances, but for the home builder, working on the main structure, no accuracy greater than 1 in 1000 was deemed necessary.

So, for the purposes of this book, three or four significant figures will generally be enough. And for a numerate life, where the aim is to be able to form a view on which numbers are important and how they compare with expectations, then three or four significant figures will generally be enough. So, 1000 is where big numbers start. More or less.

More spuriously neat ratios

The **Trans-Canada Highway** (7820 km) is
  twice the length of **Route 66** (3940 km)

Time since the earliest evidence of **farming** (11,500 years) is
  20 × time since invention of the **printing press** (576 years)

Distance from the **Earth to the Moon** (384,000 km) is
  1000 × length of the **River Thames** (386 km)

Mass of a **rhino** (2300 kg) is
  4 × mass of a **thoroughbred racehorse** (570 kg)

Time since the birth of **Charles Darwin** (208 years) is
  about 2 × time since the birth of **Alan Turing** (105 years)

Height of the **One World Trade Center** skyscraper (541 metres) is
  4 × height of the **London Eye** (135 metres)

The Second Technique: Visualisation

Paint a Picture in Your Mind

Data visualisation has become a staple of modern print and web journalism, and for good reason. By taking a dataset and capturing salient details of the data in graphic form, the artist/author connects with parts of your brain that pure text doesn't touch. I'm not going to try to impress you with cool images, nor to teach you how to go about creating such images. Instead, I want you to think about how to construct those dataviz images in your mind, as a way of understanding the big numbers you come across.

How big is a billion?

We live in a world of three spatial dimensions, but we're taught numbers in one dimension. We've already seen how hard it is to recognise even one part in a thousand on a linear scale. How can we begin to think about how big a billion is?

Well, let's start small: ants come in large and small sizes, but for this exercise I want you to imagine a tiny ant. Just 4 mm long, less than a quarter of an inch. Now, we'll take a step to something somewhat bigger than our ant. Perhaps a beetle will do? Well, if you choose the right beetle: as it turns out, a classic Volkswagen Bug measures 4.08 metres long. I'm sure you can visualise that. A column of a thousand of those tiny ants, lined up beside a VW Beetle.

Next step: what's a thousand times longer than that car? Here's a good candidate: Central Park in New York City is just the right size for us, at 4.06 km in length. Visualise it. It spans precisely 50 street blocks, which, as you know, in NYC are numbered specifically for our convenience, and Central Park stretches from 60th Street to 110th. That means each street block accounts for 80 m (including the streets themselves). So please can I ask you to visualise 20 cars, bumper to bumper, for each block? 50 of these blocks giving us 1000 Beetles stretched the length of Central Park West.

Now, we need to find something that's 1000 × longer than Central Park. And, wouldn't you know it, that's just about the East–West 'width' of Australia, 4033 km at its widest. So now we know that Australia measures 1000 Central Parks, which is equivalent to a million VW Beetles, and of course, a billion ants.

I hope those pictures are vivid in your mind. A long column of ants lined up alongside a VW Beetle. A traffic jam along Central Park West, a thousand VWs bumper to bumper. And Central Park replicated coast to coast across Australia (East to West), a thousand times. And perhaps a blurry sense of each of those Central Parks edged with cars, each car paralleled by a line of ants.

So now let's think it through. A billion is still too much to contemplate in its raw form, but we've used a little help, and a kind of a zooming-out process to break the problem down.

It's a real stretch to understand just how big a billion is. What we have done is one way to visualise it using a series of different levels. In this case, three levels, each in a thousand-to-one ratio to the next. Notice how we've made use of four new invented-on-the-spot yardsticks. Four new units of measure—four new **landmark numbers**. The ant, at 4 mm, the beetle, at 4 m, the park at 4 km, and the Oz, at 4000 km. (You may remember that the equator is 40,000 km in length—that's 10 Ozzes).

Let's take a different tack. How long is a line of a billion pennies? A British penny is 20.03 mm across. So a thousand pennies would make 20.03 m (the length of a cricket pitch) and a million in a line would make 20.03 km. A billion is 20,030 km. That's halfway around the equator, five Ozzes.

What if we were to arrange the pennies, not in a line, but spread out over an area, say on the tarmac of a large airport. How wide would they spread? If we arranged them as 40,000 rows of 25,000[20] each, that would come to 801.2 m by 500.75 m, considerably less than a square kilometre. And if we can use two dimensions, then surely we can use three? Let's pile them high, perhaps 1000 pennies per stack. A penny's thickness is just 1.5 mm, so these stacks would be

---

[20] This would not be the most efficient way to fill the tarmac. A hexagonal arrangement like a honeycomb would save some space, but this is not the place to get into that!

a mere metre and a half high, but we'd now be able to have a thousand rows of a thousand stacks, each a thousand pennies high. A square 20.03 m on a side, less than the height of a bookcase—containing a billion pennies.

Notice that two things make this visualisation different from the 'ants across Australia' one. First, we're just using one kind of yardstick, the penny coin. Second, we're making use of three dimensions, to fold up that long line of pennies into a compact shape.

Unsquaring

I was recently listening to a podcast[21] describing the fate of the stone columns that formed the old East Portico of the US Capitol building. In passing, the presenter described an area as '100 square yards'. Without stopping to think, I converted this in my head to '10 yards, squared'. 100 square yards is not a thing I can naturally visualise: 10 yards is, and a square of 10 yards on a side is just as easy.

So this is another tool for visualisation—reducing a number by folding it up into two or three dimensions. In this case, it was explicitly done since the presenter was talking about an area, but it works just as well for counts of items. A fighting force of 600 soldiers? Is that a big number? Picture them on a parade ground, in 20 (4 × 5) platoons of 30 each (3 × 10 arrays).

How many tennis balls does it take to fill St Paul's?

I came across a report on the acoustic characteristics of St Paul's Cathedral in London, and it had this snippet of information: the interior volume of the cathedral is 152,000 m³. Is that number plausible? Let's use a little bit of visualisation and some solid geometry to do a rough-and-ready cross-comparison.

A quick Google at some pictures and measurements tells me that, to a very rough approximation, the main body of St Paul's is a cuboid, roughly 50 m wide, 150 m long, and 30 m high. The famous interior Whispering Gallery is at the 30 m height and the exterior Stone Gallery around the dome is at 53 m.

---

[21] *99% Invisible*—recommended for thoughtful, articulate pieces on design and architecture, and much more along the way.

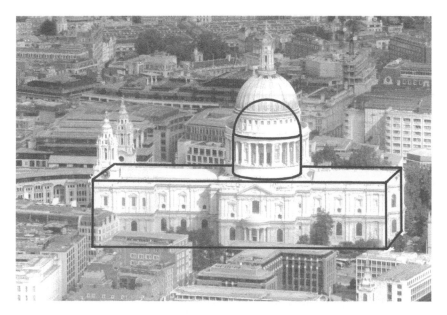

Photograph taken by Mark Fosh and reproduced here under a Creative Commons licence

Based on this, I don't think it's unreasonable to imagine a simplified shape with the following interior dimensions for the cuboid (if you pushed all the interior stonework to the edges): width 40 m × height 25 m × length of 140 m, giving a total of 140,000 m³.

St Paul's has a double dome: the inner dome sits well below the outer (and there is a brick cone between the two, which serves to strengthen the structure). The inner dome is around 30 m in diameter and it sits on a cylindrical drum. Taken together, the dome and the drum add approximately another 30 m to the interior vertical height. Working this through gives about another 18,000 m³ for the dome and drum. We've reached a total of 158,000 m³, which is enough to convince me that the figure that the acoustic engineers used (152,000 m³) is plausible.

Now for the tennis balls. If you tumble a load of balls into a container, they won't completely fill the space. If you pack them super-carefully, you can bring the proportion of space filled to around 74%, but if you just let them settle for themselves, you can expect around 65% of the space to be filled. A tennis ball of 6.8 cm diameter will have a volume of around 165 cm³, but when many of them are loosely packed, each will on average occupy a volume of around 250 cm³, roughly a cupful. This means a box with a volume of 1 m³ will hold around 4000 tennis balls (not allowing for the 'edge effect' that stops them from being so closely packed around the edges). The interior of St Paul's will therefore hold 152,000 times as many, giving a total of 608 million tennis balls.

But what if, instead of tennis balls, we used pool balls? With a diameter of 5.715 cm, their volume is very nearly 60% of that of a tennis ball. You can see where I'm going with this. A metre cubed can accommodate 6700 pool balls, and if we multiply up, we get to 1,018,400,000 pool balls to fill St Paul's. And that's one more way to visualise a billion.

PART 2

Measuring Up

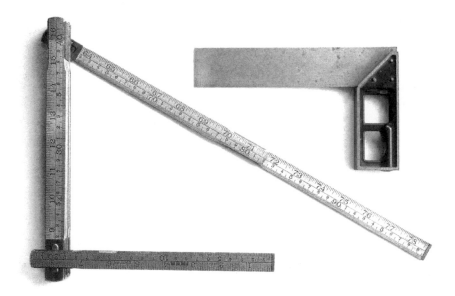

Measurement

Be the measure great or small, let it be honest in every part.   John Bright

What it means to measure

The etymology of the word 'measure' is intriguing:

> **measure (v.)**
> *c. 1300, '**to deal out by measure**,'*
> *from Old French mesurer '**measure; moderate, curb**' (12c),*
> *from Late Latin mensurare '**to measure**,'*
> *from Latin mensura*
>   *'**a measuring, a measurement; thing to measure by**,'*
> *from mensus, past participle of metiri '**to measure**,'*
> *from Proto Indo-European *me- '**to measure**'*
>
> https://www.etymonline.com

In other words, no matter how far back you trace it, 'measure' simply means 'measure'—that's how fundamental the concept is.

The title of Shakespeare's play *Measure for Measure* signifies that the severity of a punishment should match the severity of the offence. When serving food or drink, we measure out the portions. Someone who is self-controlled might talk in a measured way. When we take measures to address a problem, it implies a controlled and contained process. The common thread here is the use of a standard reference quantity to establish uniformity, balance, control, or equality.

Measuring something may involve counting, and usually does, but it is not quite the same thing as counting. Measuring, crucially, involves a unit, a reference quantity. Sailors may get their measure of rum,[1] Oliver Twist asks for more than his measure of gruel. The biblical verse Proverbs 20:10 tells us that 'differing weights and differing measures, both of them are abominable to the Lord'. Measures must be standardised, or they mean nothing.

The simplest form of measuring is indeed a form of counting: counting out how many of the standard units would be needed to make up the quantity being measured. So to know if a measurement is a big number, we need to know not only the number of units counted, but also what unit is being used. One hundred kilometres may or may not be a big number (depending on the context), but one hundred light-years is certainly a lot bigger.

When we're counting, we are dealing in integers, whole numbers, but with measurements we encounter fractions. Measuring is a way of assigning numbers (which are in fact counting numbers, even if we end up counting fractions) to continuous dimensions. The surveyor in Ancient Egypt who uses rope to lay out the foundations for a great pyramid is converting the continuous spatial dimension of length into a specific number of his selected unit. Perhaps he is lucky enough that the side of the pyramid is some exact multiple of his chosen unit, but more likely not. He needs to anticipate that when he makes his measurement, he will find that there is a remnant distance left over after all his whole units have been counted. He finds he cannot deal only in whole numbers.

Now he has a choice to make as to how to deal with this remaining quantity. The first option is to express that smaller-than-the-full-unit quantity in terms of a different, smaller unit. This is the way we might use feet and inches, pounds and ounces, dollars and cents. The other way would be to use fractions of the whole unit: one-half, two-fifths, one-eighth.[2]

A later surveyor, perhaps one of Napoleon's team of Egyptologists, might have measured the pyramid. He would have been using a relatively new system of units: the reformers who followed in the wake of the French Revolution had by then introduced the metric system, now formalised as the International System of Units (SI), to sweep away the ancient, non-decimal multiples and subdivisions of units. Our French surveyor would have written down the measurement in metres and decimal fractions of a metre (the metric system never allows a mixture of units and subunits). One way or another, though, measurement needs to deal with fractional units.

---

[1] A navy tot of rum was an eighth of an imperial pint, around 70 millilitres, that's two 35 ml shots.

[2] In fact, the Ancient Egyptians would not have used two-fifths. They preferred to work in terms of unit fractions (one-over-something) or sums of unit fractions. Two-fifths would have become $1/3 + 1/15$.

Measuring, then, is a fancier form of counting: it is counting with units attached, and needs to accommodate the fact that the numbers may be fractional, or that there may be sub-units involved. So we'll start by looking at some of the ways we measure that most fundamental of quantities, distance. There'll be plenty of **landmark numbers** to find, and plenty of opportunity for **visualisation**. And there's one measuring tool that's available to all of us: the human body.

About the Size of It

Numbers to Quantify the Space We Live in

Measure what is measurable, and make measurable what is not so.   **Galileo Galilei**

Which of these is the longest?
- ☐ A London bus
- ☐ Estimated length of Tyrannosaurus Rex
- ☐ Distance a kangaroo can jump
- ☐ A T-65 X-Wing starfighter in *Star Wars*

The Long and the Short and the Tall

And this is the fashion which thou shalt make it of: The length of the ark shall be three hundred cubits, the breadth of it fifty cubits, and the height of it thirty cubits.

King James Bible: Genesis 6:15, specification for Noah's Ark

—The Equator is 40,000 km in length. *Is that a big number?*

—The Empire State Building is 381 m tall. *Is that a big number?*

—The Zambezi River is 2574 km long. *Is that a big number?*

Before we digitised everything, almost all measuring instruments, rulers, clocks, voltmeters, thermometers, scales, even protractors, were analogue devices that converted the quantity under investigation into a linear equivalent. This meant that we would make a measurement by reading the position of a hand, a needle, or a mercury column against a calibrated scale either on a straight edge or the arc of a circle. All measurements were converted to linear distances. So it makes sense to start by discussing that most basic, most fundamental of measures: distance.

Made to measure

Nothing could be more natural than to measure things with those measuring sticks that are most readily available: body parts. It's easy to see how the very first crude measurements would have been made with whatever was close at hand, or even the hand itself.

For small lengths the Greeks used fingers (*daktyloi*[3]) and feet (*podes*) for measurements, and there were 12 fingers to the foot, and the feet were roughly the same size as the imperial foot we use today. The Romans borrowed the fingers and feet from the Greeks, and brought them to Britain. The fingers became

---

[3] And yes, there is an etymological connection with the flying dinosaur, the pterodactyl, the 'wing-fingered' creature.

uncia in Rome, which we in English started calling inches, but which all the rest of Europe realised were really fingers, or more precisely, thumbs.[4]

Horses are measured in hands to this day, and the cubit, used in the specification for Noah's Ark, is the length of the forearm and hand, sometimes considered to be roughly a foot and a half.[5] And I can still remember my father asking for 'two fingers' of whisky.

'Fathoms' are used to measure water depth: and the fathom is another unit name taken from the human body, another anthropic unit. It comes from a Proto-Germanic word *fathmaz* meaning embrace, and hence 'arms outstretched' and is around 6 feet. The old French unit named the 'toise' measures the same span, and that word comes ultimately from the Latin word for 'outstretched' (and has etymological connections with 'tent' and 'tension').

Of course, to properly turn these informal measures into established systems, they needed to be standardised. So, even though the names referred to body parts (and no doubt actual body parts were used for casual measurements), official standards, usually in the form of physical measuring sticks, were needed. But the folk names live on as testament to the origins of these units. This use of the body as a measuring stick still lingers in our language when we talk of a 'rule of thumb'. And, idiomatically, the narrowest of margins is a 'hairsbreadth'.

When thumbs aren't quite big enough, we look for a 'yardstick'. The origin of the word yard is unclear (some relate it to the girth around the waist, which really does seem a wilfully unreliable measure, even worse than fingers and feet).[6] However, the yard has a central position in the imperial and other ways of measuring: most systems of units seem to have a close equivalent to the yard. And the word 'yardstick', once physical, now means any practical reference for making measurements.[7]

---

[4] The word 'inch' derives from the Latin *uncia* (also the source of 'ounce') and means one-twelfth; but in many other European languages the word for 'inch' means 'thumb': In French, *pouce* = inch/thumb; in Dutch, *duim* = inch/thumb; in Swedish, *tum* = inch, *tumme* = thumb; and in Czech, *palec* = inch/thumb.

[5] The 'ell' was originally the same as the cubit (and the 'elbow' was the joint that set its limit), but at some stage it more than doubled, so that an English ell, as used for measuring cloth, became 45 inches, while a Scottish ell was 37 inches.

[6] A more likely derivation of the yard could be as half a fathom, the measurement from the central axis of the body to the end of an outstretched arm.

[7] And that reminds me of the word 'benchmark', a lovely example of a widely used abstract concept, which so plainly derives from having an actual mark on an actual workbench.

Out and about

When the Romans went marching, they measured distances in terms of how many steps they took. 125 paces (double steps) was a 'stadium'—and yes, our name for a sports arena comes from the same root.[8] 8 stadia, or 1000 double steps, was called a *mille* (plural *milia*), and this has become our 'mile'. The Romans' double step (a pace or *passus*) was 5 feet, and so their mile was 5000 feet. Dividing this number by three, we get to around 1667 yards, not so different from the modern definition of the mile at 1760 yards.[9]

In Rome, one and a half miles became a *leuga*, or league. However, at some point in its translation to mediaeval measures, this number was doubled, and so the folk definition of a league became the distance a person could walk in an hour, generally taken as 3 miles. Seven-league boots, then, could cover 21 miles with every step.

A furlong ('furrow's length') was the length of furrow that a team of oxen could plough without resting, and is more or less equal to the length of the Roman 'stadium'. Nowadays, the use of furlongs is restricted mostly to measuring horse races in English-speaking countries, although some American cities still retain grid systems based on the furlong.[10] A furlong is 220 yards, just a little more than 201 metres.

In 1620, Edmund Gunter started using, for survey work, a physical chain equal in length to $^1/_{10}$ of a furlong, 22 yards, and this gave rise to a unit of measurement used almost exclusively in land surveys, called the 'chain'. A rectangle of land one furlong in length and one chain in width is an acre of land. And that makes a furlong squared equal to 10 acres.

The Ancient Egyptians used their hands for measuring in a variety of ways, and had units called the finger, the palm (four fingers), the hand (five fingers), and the fist (six fingers). For bigger measurements, they used spans, cubits, poles, and rods in many variants, and a large unit called the *iteru*, a 'river', of 20,000 cubits (equating to around 10.5 km). Archaeologists have found tangible evidence of these systems of measurement in the form of cubit rods marked

[8] The 'stadium' distance took its name from a running race, the *stadion* in Greek games, which took its name from the venue for the race, also a *stadion*. The original stadion at Olympia was around 190 metres long.

[9] The Roman 'stadium' was therefore 625 'pedes', while the Greek 'stadion' was 600 'podes'. 'Pedes' and 'podes' both mean 'feet'.

[10] The city block systems in Chicago and Salt Lake City are based on furlong measurements.

with subdivisions into palms, hands, and fingers. For the larger measurements, the Egyptians used ropes with knots at specific intervals.[11]

Traditionally a Chinese foot—*chi*—was around 32 centimetres long. Five of these made a *bu*, a pace, which was around 6 (Western) feet, and so similar in length to the Roman pace. The modern definition of the *chi* has been harmonised with the metric system and is now exactly one third of a metre long. *Chi* (both old and new style) may be divided into 10 units called *cun*, the 'Chinese inch'. Traditionally, this was the width of the thumb, measured at the knuckle.

In Ancient China, the length of a bolt of silk was commonly used as a standard unit of measure, 12 metres long. Interestingly, such a bolt of plain-weave tabby[12] silk was also widely used as a unit of currency.

In 2008, Barbara Wilson and Maria Jorge deciphered a system of measures used by the Aztecs for land surveying and calculation of areas of land. The base unit is called the *Tlalcuahuitl* (T), and was a land rod between 2.3 and 2.5 metres long. The divisions of the *Tlalcuahuitl* form a curious set:

- the arrow: $^1/_2$ T = 1.25 metres
- the arm: $^1/_3$ T = 0.83 metres
- the bone: $^1/_5$ T = 0.5 metres
- the heart: $^2/_5$ T = 1.0 metres
- the hand: $^3/_5$ T = 1.5 metres

Once again, the human body is called on to perform measuring duties.

Metre for measure

These measures based on human body parts and activities served well in the limited settings in which they arose, but were ill-suited to the requirements of the natural philosophers of the Enlightenment, who needed a more universal system, capable of framing, measuring, and coordinating a wider world of exploration and experimentation.

The French Revolution provided an opportunity for all sorts of overturning of traditions, not least in the world of measurement. And so the metre was born, and with it, the whole metric system.

[11] A loop of rope divided into 12 equal sections could be pulled out into a triangle with sides of length 3, 4, and 5 sections, and this would form a right-angled triangle. Thank you, Pythagoras.

[12] 'Tabby' refers to the wavy or moiré pattern woven into the silk and indeed shares a derivation with the tabby cat.

No longer based on human-scale concerns, the metre was born to be global. Literally: it was first defined as one ten millionth part of the distance between the North Pole and the equator, the favoured line of longitude naturally being the one going through Paris. Happily, this measure was quite close to the yard in size. So, even though it was defined using a quarter of the world as a base, the metre remains a human-scale measure.

Landmark numbers
The distance from pole to equator along a line of longitude is 10,000 km, or 10 million metres.

That 'meridional' definition of the metre no longer holds. From 1889 to 1960, the standard for the metre was established by reference to a physical prototype metre bar, and since 1960 it has been defined as:

1,650,763.73 wavelengths of the orange-red emission line in the electromagnetic spectrum of the krypton-86 atom in a vacuum.

The so-called metric system as a whole was very widely adopted (helped by the fact that it was based on multiplying and dividing by 10 rather than 12s, 14s, 16s or other eccentric multiples). And although the metric system has so far failed to fully conquer all corners of the world,[13] it remains the only real choice for a world-wide standard.[14]

Measuring the things around us

All areas of human experience are measured and counted, but few as visibly as sports. Sports are ruled and regulated by numbers. Playing fields and courts are laid out, scores are counted, records are broken—all of which demand a basic level of numeracy.

---

[13] The USA remains one of only three countries in the world not to have adopted the metric system—the others being Burma (Myanmar) and Liberia. In the UK, there is an ambivalence as to the use of metric measures, and characteristically there is an uncomfortable muddled compromise in effect, with specific exemptions permitting the use of imperial measures, for example miles on roadside signs, and pints for selling beer and milk.

[14] For that reason, in the rest of this book, we will, where sensible, default to the metric system, for distance favouring millimetres, metres, and kilometres according to the size of what we are measuring, and we'll use the standard abbreviations 'mm', 'm', and 'km' for these. But, however sensible the metric system is, the eccentric older units have an undeniable charm, and we'll pay due respect to those from time to time as well.

Sports: heights

How tall is a basketball player? In the USA men's basketball team for the 2016 Olympics, all but two members were a little over 2 metres in height. The hoop itself is just about 3 metres in height.

The crossbar of a rugby football[15] goalpost is also 3 metres, as is the bar of an American football goalpost, just the same as the height of a basketball hoop, while the crossbar of a 'soccer' football goal is only 2.44 metres high. This also happens to be a smidgen under the world high jump record set way back in 1993. So the best high jumper can jump over a soccer goalmouth—but only just!

> **Landmark number**
> - A basketball hoop
> - American football goals
> - Rugby goals
> They are all 3 metres in height. That's a bit more than half again the height of a very tall person.

The world pole vault record is two and a half times as much as the high jump record (2.45 metres). At around 6.16 metres, that's a bit over twice the height of a basketball hoop. Interestingly, the women's pole vault record (5.06 metres) is also around two and a half times the women's high jump record (2.09 metres), both of them approximately five-sixths of the men's records.

Ice hockey goalmouths are dinky, just 1.2 metres high, while Olympic hurdles are just over 1 metre in height, 1.067 metres to be precise.

In Olympic diving, the divers use either a springboard at 3 metres or a high platform at a neatly memorable 10 metres in height.

> **Landmark number** The high platform in Olympic diving is at 10 metres, about the height of three storeys.

---

[15] What do we mean by 'football'? Wikipedia takes the easy way out, defining it as 'whichever form of football is the most popular in the regional context in which the word appears'. I'll use rugby (Union or League specified if the distinction is relevant), soccer (forgive me!), American football, and Australian rules football as appropriate.

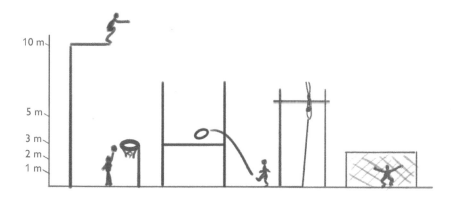

The uneven bars in women's gymnastics are set at 2.5 metres and 1.7 metres, while the men's high bar apparatus is 2.75 metres high.

Sports: distances

Starting small: a table tennis table is 2.74 metres (9 feet) long. A regulation pool table can be one of two standard lengths: 2.74 m = 9 feet, the same as for table tennis, or the smaller size of 2.44 metres = 8 feet.

A ten-pin bowling lane is 18.29 metres from foul line to the forwardmost pin (and 1.05 metres wide).

A tennis court is 23.77 metres long (78 feet), and the height of a tennis net is 0.914 metre (3 feet), traditionally measured using two tennis rackets, one contributing its length and the other its width to make an impromptu measuring stick.

Basketball courts are 28 metres long, a bit more than 10 times the length of one of the larger pool tables, and 15 metres wide.

A cricket pitch is 20.12 metres long from one set of stumps to the other (naturally the imperial measure came first, this is 22 yards and therefore is 1 chain long, or one tenth of a furlong), although the specially tended strip has an extra 4 feet beyond the stumps on either side, making the full length 22.56 m). The cricket **field** is whatever shape and size the field (or village green) happens to be.

Likewise, the baseball diamond is set in a field of arbitrary size. The sides of the diamond are 27.43 metres long (but, as with the cricket pitch, the imperial measure of 30 yards is neater and more memorable). This is very close to the length of a basketball court, and is equal to the length of ten large pool tables end-to-end.

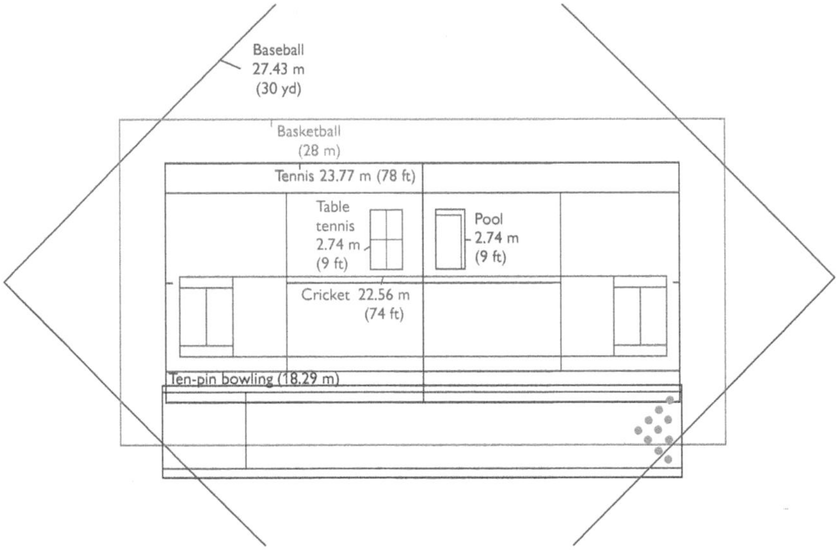

Baseball
27.43 m
(30 yd)

Basketball
(28 m)

Tennis 23.77 m (78 ft)

Table
tennis
2.74 m
(9 ft)

Pool
2.74 m
(9 ft)

Cricket 22.56 m
(74 ft)

Ten-pin bowling (18.29 m)

How big is a football field?

For soccer, the regulations are rather flexible: anything between 90 and 120 metres long and between 45 and 90 metres in width is acceptable,[16] but in fact most Premier League pitches in the UK are around 105×70 metres. American football fields must be 110 metres long and 48.76 metres in width, giving an aspect ratio that is rather more elongated than a soccer pitch—in fact, it's more than 2:1. It's a curiosity that the length of an American football field is nearly equal to the combined distance between the four bases of a baseball diamond.

Rugby fields have 100 metres between each set of goalposts, but have 'in-goal areas' beyond the goalposts on each end. Rugby league pitches are 68 metres wide, while for the rugby union code, they must be a minimum of 70 metres wide.

Gaelic football is played on a larger pitch, 130–145 metres long and 80–90 metres wide. Australian rules football fields are oval, and have no regulated size, but are typically around 160 metres long and 120 metres wide. So they are considerably larger than most other football fields.

---

[16] Intriguingly, this means that a 90 metre square pitch would be legal.

Landmark number
Length of a football field = Choose your own preference:
- Soccer: 105 m
- American football: 110 m
- Rugby: 100 m + in-goal areas
- Gaelic football: 130–145 m
- Australian rules football: typically around 160 m

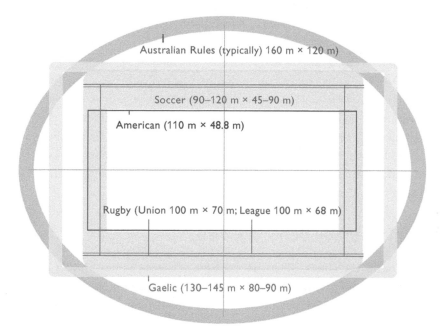

Australian Rules (typically) 160 m × 120 m)

Soccer (90–120 m × 45–90 m)

American (110 m × 48.8 m)

Rugby (Union 100 m × 70 m; League 100 m × 68 m)

Gaelic (130–145 m × 80–90 m)

So a school sports field will generally be at least 100 metres in length, not only to allow the desired kind of football to be played, but also to accommodate running races of a suitable length.

The Ancient Greeks based one of their units of length, the stadion, on the length of a running race. It was named for the venue where the race was run (hence, our word 'stadium', via the Romans), and was more or less 170–200 metres long (it was always 600 feet, but the definition of the length of the foot was not always agreed on), and it's generally thought that this was the length of the shortest sprint that was run competitively. These days, of course, sprinters will run their shortest races over 100 metres, and successively longer races follow a doubling pattern: 200 metres (no coincidence that this is pretty close to a furlong), 400 metres, and 800 metres.

The next running distance is rounded to 1500 metres, although 1600 metres would more closely approximate the one-mile race[17] (which would come to 1609.3 metres). Long-distance races are typically run over 3000 metres, 5000 metres, and 10,000 metres. And finally, to bring us right back to the Greeks, the marathon race commemorating the run of the messenger Philippides[18] from the battle of Marathon to Athens is now formalised as a distance of 42.2 kilometres. And that's just about 400 football pitches, and was run by the winner of the men's event at the Rio Olympics in 2016 in just under 2 hours 9 minutes, and by the winner of the women's event in 2 hours 24 minutes.

The story of Philippides running to Athens is probably not true. The Greek historian Herodotus does, however, record a run made to Sparta to summon aid before the battle of Marathon. The distance from Athens is more like 250 km and reportedly took Philippides two days. In 1982, a group of four RAF friends decided to recreate this run and managed the distance in 36 hours. It has since become an annual event called the Spartathon. The fastest time stands at 20 hours and 25 minutes.

How far can you throw that thing?

Thinking of the Greeks in battle, how far can a spear be thrown? There's no reliable evidence in the historical record, but we do know that in 1984 Uwe Hohn threw a modern javelin a distance of 104.8 metres (almost exactly a soccer field's length), a record that may stand for some time, as the technical specifications for a competition javelin have since changed. The men's javelin event at the 2016 Olympics was won by Thomas Röhler with a throw of 90.3 metres.

A discus has been thrown 74.08 metres, and a 7.26 kilogram shot has been thrown 23.12 metres. The longest recorded flight for an arrow shot from a bow is reported as 'nearly 500 metres'.

---

[17] The four-minute mile was and is a wonderful landmark number. Even if the current record of 3:43.13 held by Hicham El Guerrouj of Morocco breaks it by more than a quarter minute, it's neatly memorable to think of a mile being as much as an elite athlete can run in 4 minutes, a distance you or I would take 20 minutes to walk.

[18] Or Pheidippides, or Phidippides: take your pick.

Landmark number
- Javelin throw: a bit under 100 m
- Discus throw: three-quarters of the javelin throw
- Shot put: one-quarter of the javelin throw
- Bow and arrow: five times as long as the javelin throw = nearly 500 metres

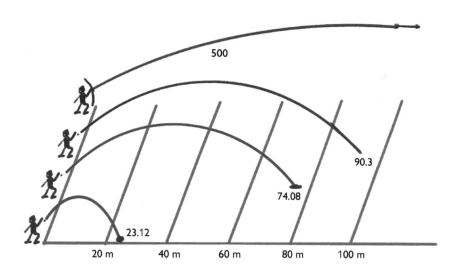

An (American) football can be thrown 80 metres. A cricket ball has been thrown 125 metres, and a throw of 135 metres is achievable with a baseball. The longest golf drive ever recorded was 471 metres.

Sports: equipment

Baseball and cricket bats are roughly similar in regulated length: a minimum of 1.067 metres for a baseball bat; a maximum of 0.965 metres for a cricket bat. Either would serve, at a pinch, as crude yardsticks.

A table tennis ball is 40 mm in diameter, a golf ball is 43.7 mm, a pool ball is 57.2 mm, a tennis ball around 67 mm and a baseball is 73.7 mm, while a basketball is at most 241.6 mm in diameter, needing to pass through a hoop of 457 mm, a bit less than twice the width of the basketball itself.

The Long versus the Tall

Consider the Earth's average radius of 6370 km—centre to surface. Consider the height of Mount Everest—8.8 km above sea level, and the depth of the Marianas Trench—11 km below sea level. Taken together, this means that there is a range of a little less than 20 km vertical difference between Earth's highest peak and its lowest trough. That's just 1 part in 320. The equivalent on a table tennis ball, with a 20 mm radius, would be a mere $^1/_{16}$ of a millimetre, barely discernible.

When we are travelling around the world, most of the distance we cover is horizontal distance: tens of thousands of kilometres. On the other hand, the greatest possible vertical distance we can climb or fall while staying on the surface of the Earth is not even 20 kilometres. So, even though heights and distances may both be measures of linear spatial extent, the horizontal–vertical distinction makes for poor comparisons. On this planet, flattened by gravity, vertical heights are just so much smaller than horizontal distances.[19]

Buildings and other structures

The highest a pole vaulter has yet vaulted is just over 6 metres, which is a little shy of 2 storeys of a building. High-divers dive from a height of around 3 storeys high. And when we humans want to vaunt wealth and show off status, we tend to build towers—towers of great height.

> **Rule of thumb**
> A skyscraper's height is around three and a half metres per floor, plus ten metres. Try using this rule on some of the buildings described below.

In San Gimignano in Tuscany in the twelfth century, there arose a passion for showing off one's status by building towers. There was little land available within the walls of the hilltop town—no room for palaces—and so towers became the thing. Every family of consequence had their tower, each tried to trump their neighbour. At the height of the craze, there were 72 in all, and they rose up to 70 metres tall, the equivalent today of 18-storey buildings (and more

---

[19] For that reason, in this book and on the companion website, we try only to compare verticals with verticals or with orientation-free reference lengths (for example, the length of a baseball bat), and likewise for horizontals. Once we escape the Earth's gravity, of course, the problem goes away, pretty much. There is no 'up' in space.

than 11 times a world record pole vault). Now just 14 of them remain, but they nonetheless make for a remarkable skyline for a small hilltop town.

The first building to be called a skyscraper,[20] the Home Insurance Building in Chicago, was built in 1884. It was built with 10 floors at 42 metres, but two extra floors were added in 1890, taking it to 55 metres in height, still considerably shorter than the towers of San Gimignano. But then it was all practical, useful space.

Perhaps the world's most iconic skyscraper for much of the twentieth century was the Empire State Building. Still matched only by the Statue of Liberty (a height of 93 metres from ground to tip of torch[21]) as a symbol of New York City, its 102 floors take it to 381 metres,[22] some seven times the Home Insurance Building's height.

> **Landmark number** The Empire State Building is 102 floors high, and about 381 metres tall.
>
> (The Statue of Liberty is about one-quarter of that height.)

The Empire State Building's status as the world's tallest building was taken from it by the 417 metre World Trade Centre, so shockingly and tragically brought down in 2001. In its place there now rises a new One World Trade Centre, the tallest building in the Western Hemisphere. It is 541 metres high—although without its narrow spire, its roof height is 407 metres. Its height is said to be 104 standard floors, although there are only 94 actual floors above ground.

At the time of writing, the Burj Khalifa in Dubai is the tallest skyscraper in the world: its architectural height is 828 metres, although its highest floor is only at 585 metres—there are 46 spire levels. Architects call the unusable height of these very tall buildings the vanity height. In some cases, as with the Burj Khalifa, the vanity height appears very much in harmony with the overall design. In other cases, the addition of a spire looks gratuitous, and one is inclined to suspect vanity or rivalry may indeed be the motivation.

> **Landmark number** The Burj Khalifa is 828 metres high.

---

[20] Imagine you had never known the word skyscraper and were hearing it for the first time. Isn't it the most poetic of terms?

[21] The copper statue itself is 46 metres high, a little under half of the overall height.

[22] This is called its architectural height, including spires, but excluding antennae or flagpoles. The use of roof height as a measure has mostly fallen away as so many tall buildings now have no definitive roofline to measure to.

But if we decided not to count vanity height, then most of the height of St Paul's Cathedral in London would have to be discounted, and that would never do. St Paul's, at 111 metres, reigned as the tallest building in London for 252 years, from 1710 to 1962, and was displaced from this status only by the Post Office Tower, which reached 183 metres on completion in 1964. Today, the tallest building in London is The Shard of Light, with 95 storeys rising to 310 metres.

That's still not as high as the iconic Eiffel Tower, the landmark of Parisian landmarks, which is 324 metres high, and was initially regarded as a temporary construction, built for the 1889 World's Fair.

Landmark number The Eiffel Tower is 324 metres tall.

The Burj Khalifa will likely be overtaken in 2020 by the Jeddah Tower in Saudi Arabia, which will be the first building over a kilometre tall, at 1008 metres, though its height too will be exceeded if plans come to fruition to build a 1054 metre tower in Azerbaijan. That would make it 12% the height of Mount Everest, measuring Everest from sea level.

Routes and roads

Let's come down to Earth from those lofty aeries and do a little mental arithmetic: if a league is a measure of how long a person walks in an hour (at around 5 km/h), how far could a person walk in a day? We can imagine that a good day's walking would be around 8 hours, and that would mean we would notch up perhaps 40 km of distance travelled.

How far is that? Well, it is:

- 10 times the length of New York's Central Park
- ¹⁄₂₀ the length of the Indianapolis 500 motor race (500 miles)
- ¹⁄₁₀₀ the East–West width of Australia
- ¹⁄₂₀₀ the North–South length of Africa
- ¹⁄₁₀₀₀ the length of the equator

Landmark number It would take 1000 days to walk around the Earth.

How much faster would it be if you were driving a car? Around 20 times faster (at 100 km/h). So a good day's driving (again, 8 hours) would cover 800 km.

At that rate it would take around 50 days to drive around the Earth's equator. Jules Verne's Phileas Fogg did well to cover it in 80 days—he would have had to cover 500 km a day, but then he often used 24-hour transportation.

> **Landmark number** It would take 50 days to drive around the Earth.

What about flying on a commercial flight? That would be around 800 km/h—so 8 times faster again. If you were flying only 8 hours a day, it would take 6 and a bit days. But, assuming you could make perfect connections, you could accomplish the trip in around 2½ days.[23] In 1980 a military B-52, refuelling in mid-air, did indeed fly around the world in 42 hours and 23 minutes.[24]

> **Landmark number** You can fly around the world in 2½ days.

The International Space Station completes the trip at an altitude of 400 km in just 92 minutes. But Shakespeare's Puck can outdo that. He can 'put a girdle round about the Earth in forty minutes', more than twice as fast as the ISS. That speed has him travelling at a very neat 1000 kilometres per minute. Maybe we should coin a new unit, the Puck, equal to a megametre per minute. Has a ring to it, don't you think?

> **Rule of thumb** Flying time on a commercial flight between two cities.
>
> Take the approximate distance in thousands of kilometres, add a fifth, and add another half hour for take-off and landing, and you have the flight time in hours. Try it! London Heathrow to New York's JFK is around 5.5 thousand kilometres. Add a fifth to get 6.6 and then a half (more or less) to get 7 hours.

Crossing continents

It's one thing to imagine walking or whizzing around the equator, as if there were an actual physical path or roadway girdling the Earth, but it's another to

---

[23] I checked: fly New Zealand Air, London to Auckland, stopping at LAX (26 hours). You'll need to spend 2½ hours on the ground at Auckland, then take a British Airways flight via Sydney and Singapore, which takes 31½ hours, making exactly 60 hours in total.

[24] In passing, notice how natural it is to think about distances measured in terms of time. In past times, a town might be described as being 2 days' walk away, while nowadays you might say a shopping mall is 'a 20-minute drive' away. This way of expressing distances using time even carries through to the scientific world in the form of a 'light-year'.

contemplate actual long journeys across the face of the Earth. The names of the great journeys resonate with romance and adventure: 'The Trans-Siberian Express', 'Cape to Cairo', 'The Silk Road', 'The Orient Express', even 'Land's End to John O'Groats'.

But let's begin with the one famed in mid-20th-century song: Route 66. As the song tells us: 'It goes from Chicago to LA/More than two thousand miles along the way'. Is that right? Well, Route 66 sadly no longer exists as a continuous maintained route, but yes, it's more than 2000 miles—in fact it is around 2450 miles long or just shy of 4000 km. If the roads were good, and we could travel at our hypothesised 800 km a day, that would be 5 days' travel. In practice, the time recommended to tourists is two weeks, allowing for slower roads, and time for sightseeing.

> **Landmark number** Route 66 is (or was) 4000 km 'along the way'.

Should you want to drive across the USA somewhat faster on Interstate Highways, the I-90 may be the one for you. It covers 4860 km from coast to coast, Boston to Seattle. In good weather, 6 days at 810 km a day is about what you would need.

> **Landmark number** Boston to Seattle—just a bit under 5000 km, taking 6 days.

That's between 3 and 4 times the distance needed for the longest possible journey in Great Britain, from Land's End in Cornwall through to John O'Groats in Caithness in the North-East of Scotland. As the crow flies, that is just 970 km, but on roads it's more like 1350 km.[25] Some of the roads would be fast, some would not. 16 hours of driving means you should allocate 2 days for that trip.

> **Landmark number** As the crow flies, the distance across the island of Great Britain from Land's End to John O'Groats is 970 km.

The longest railway line in the world is the Trans-Siberian Railway running from Moscow to Vladivostok. It's 9290 km long. If stretched straight, that would be a little less than a quarter of a great circle around the Earth. The trains on that route can make around 900 km in a day in the best conditions—so plan on 11 days for the trip. However, plans in place to speed up those trains and reduce

---

[25] For fans of fractions, that's $5/18$ of the I-90 route.

delays should increase the daily distance travelled to 1500 km per day, which would result in a 7-day trip.

The original Orient Express started in Paris and ended in what was then Constantinople (now Istanbul), a trip of around 2800 km. You can still make that trip: it takes 7 days, costs almost $20,000 and runs once a year.

And the legendary Silk Road linking China with Europe? Well, that was always a network rather than a single road, but we can calculate that from Xi'an in China to Venice in Italy, as the crow flies, would be around 7800 km, and on the ground would probably have exceeded 10,000 km. A round trip is estimated to have taken two years. Now a New Silk Road is being planned—a Chinese initiative to improve trading routes to Western Asia and Europe.

Speaking of crows flying: Australia is almost exactly 4000 km along its longest dimension, its width East to West (and that's one-tenth of the length of the equator). Africa is twice that size, 8000 km, along its longest dimension, which runs North to South.[26] South America measures 7150 km top to tip, North to South, and North America is 8600 km North to South.

Eurasia is the largest landmass on Earth, but things get complicated when you try to measure it tip-to-tip, as the crow flies. For if a crow wanted to fly the shortest possible route from the westernmost tip of mainland Eurasia in Portugal to the far eastern tip in Russia, it would in fact be better not to fly over Eurasia at all, but to take a polar route over North America. Such a route, following a great circle, would pass close to the North Pole, and would be heading more West than East. Both the westernmost point of Eurasia, and its easternmost point lie in the Western Hemisphere, even though the bulk of the continent lies in the Eastern Hemisphere.

If the crow felt that this wasn't quite in the spirit of things, and instead headed eastwards from Portugal, breaking his journey at Istanbul, famously the bridge between Europe and Asia, he would fly 3200 km in Europe and 8000 km in Asia for a total continental trip of 11,200 km.

Rivers

Before we leave the Earth-measuring ('geometry') of this chapter, a quick look at the rivers. The longest river, the Amazon, is only 8 km short of a memorable 7000 km. The Nile is 2% shorter at about 6850 km. The longest river in Asia

---

[26] The famous 'Cape to Cairo' trip would be less, in fact 7248 km as the crow flies, as Cairo is far from being the most northerly point of Africa. To make an actual road trip from Cape to Cairo would be more than 12,400 km, though.

comes next, the Yangtze at 6300 km, followed narrowly by the Mississippi–Missouri at 6275 km.

> **Landmark number** The Amazon River is 7000 km long.

And then comes the river that I had never heard of until I started researching this topic. It's the Yenisei, at 5500 km long. Rising in Mongolia, it flows through northern Asia, in Russia for 97% of its length.

5500 km. Is that a big number? Well, it's about as far as from London to New York, it's half our crow's journey over Eurasia, and it's $^1/_{25}$ of the diameter of Jupiter. We won't go into space quite yet, but we will return to distance measuring in a later chapter on Astronomical Numbers.

The Lego world

How tall would a Lego version of the Empire State Building be? In fact the Lego company market various architectural ranges that have versions of that landmark building. But for this book, we like things to be calibrated to a human scale, and in the Lego world, the humans are the miniature figures or minifigs.

So, to establish a scale, we need to decide how tall a human the standard minifig should represent, and what scale would be consistent with this height. A scale is a **ratio** that relates the model to the real thing, and using ratios is an important way of understanding big numbers. It's one of our five techniques.

Now the first thing to notice is that a minifig's proportions are wrong, so we can't be too purist about this. Minifigs are short and squat. And the second thing is that we want a practical scale. So roundish numbers will be preferred, thank you.

A minifig is 40 mm tall, which is to say 4 Lego bricks tall. The most obvious practical conversion would be 1 mm to 50 mm, but then we'd end up with a giant of a person (2 metres tall and ridiculously wide). Instead, I propose taking that down 20% and using a conversion of 1 mm to 40 mm. That conversion makes a minifig 1.6 metre tall, or in imperial measures, 5 foot 3 inches. Seems fair?

So a standard Lego brick, with a height of 10 mm,[27] represents 400 mm in our scaled-down world. A doorway of 2 metres will need to be 5 bricks high. Seems good to me.

What about horizontal dimensions? The stud-to-stud distance on a Lego brick is 8 mm, which, using our conversion, would represent 320 mm, or just over a foot. More or less, we can reckon that three studs would make a metre. This scale would mean that a standard 2 × 4 Lego brick represents a breezeblock something like 0.64 metre wide × 1.28 metre long × 0.4 metre high.

So, let's return to the Lego Empire State Building. How tall would that be? The real thing is 381 metres—so the Lego version would need to be $^1/_{40}$ of that—9.53 metres high. That would be 953 layers of bricks tall. Each floor would need on average between 9 and 10 bricks' height.

Rather wonderfully, someone has actually decided to do this! 'Kevin F' has made a start on a model at this scale and has documented his project on YouTube. The project is in its early stages and clearly making slow progress— but good luck, Kevin!

How far is 1000 km?

1 m Maximum length of a cricket bat (965 mm)
Maximum length of a baseball bat (1.067 m)

2 m Length of a king-size bed (1.98 m)
Wingspan of a wandering albatross (3.1 m)

5 m Length of a classic Ford Mustang (first generation) car (4.61 m)
Length of the longest recorded reticulated python (6.5 m)

10 m World long jump record (1991) (8.95 m)
Length of a London bus (11.23 m)

20 m Men's Olympic triple jump record (18.09 m)
Length of a cricket pitch (20.12 m)

---

[27] Actually, the specifications say 9.6 mm, but, when stacked, on average each layer should add 10 mm.

50 m	Length of the *Millennium Falcon* spaceship in the *Star Wars* movies (34.8 m) Length of an Airbus A380 (72.7 m)
100 m	Length of the *Cutty Sark* sailing ship (85.4 m) World record javelin throw (104.8 m)
200 m	Length of the Pont Neuf bridge over the Seine (232 m) Length of the *Titanic* (269 m)
500 m	World record longest flight of an arrow from a bow (2010) (484 m) Length of the Lincoln Memorial Reflecting pool in Washington DC (618 m)
1 km	Length of Tiananmen Square in Beijing (880 m)
2 km	Length of the Champs-Élysées (1.9 km) Length of the Epsom Derby horse race (2.4 km)
5 km	Length of Copacabana Beach in Rio de Janeiro (4 km) Length of the Oxford–Cambridge Boat Race (6.8 km)
10 km	Length of the longest canal tunnel (Rove Tunnel in France) (7.12 km) Length of Marina Beach in Chennai, India (13 km)
20 km	Length of Manhattan Island (21.6 km) Narrowest width of the English Channel (32.3 km)
50 km	Length of a marathon race (42.2 km) Full length of Broadway (53 km)
100 km	Length of the Kiel Canal (98 km) Diameter of the Death Star in the first *Star Wars* movie (120 km)
200 km	Length of the Suez Canal (193.3 km) Length of the British Grand Prix Formula 1 motor race (306.3 km)
500 km	Length of the River Thames (386 km) Distance from London to Edinburgh (535 km)
1000 km	Direct distance from Land's End to John O'Groats (970 km) Length of mainland Italy (1185 km)
2000 km	Length of the Tigris River (1950 km) Length of the Zambezi River (2574 km)
5000 km	Diameter of Mercury (4880 km) Length of the Yellow River (Huang He) in China (5460 km)

10,000 km   Length of the 2016 Dakar Rally (9240 km)
            Length of the Trans-Siberian Highway: St Petersburg to
                Vladivostok (11,000 km)

20,000 km   Range of an Airbus A380 (15,200 km)
            Length of the equator (40,100 km)

How high is 100 metres?

2 m   Women's Olympic high jump record = 2.06 m

5 m   Women's Olympic pole vault record = 5.05 m

10 m   Height of a high-diving platform = 10 m

20 m   Height of All-Terrain Armored Transport (AT-AT) in *Star Wars*
            movies = 22.5 m

50 m   Height of the main mast of the *Cutty Sark* sailing ship = 47 m
        Height of Niagara Falls = 57 m

100 m   Length of the *Saturn V* Apollo Program launch rocket = 110.6 m
        Height of St Paul's Cathedral in London = 111 m

200 m   Height of Sagrada Familia Basilica in Barcelona = 170 m
        Height of The Shard skyscraper (London) = 310 m

500 m   Height of the One World Trade Center skyscraper in New York
            City = 541 m

1 km   Height of the Angel Falls (Venezuela) = 979 m

2 km   Depth of the Mponeng Gold Mine (South Africa) = 3.9 km

5 km   Height of Mount Kilimanjaro = 5.89 km

10 km   Height of Mount Everest = 8.85 km
        Depth of the Marianas Trench = 10.99 km

Who'd have thought that (to within 2%)...

The **Great Wall of China** (8850 km) is
  about as long as the **USA–Canada border** (8890 km)

The height of **The Shard** skyscraper (London) (310 m) is
  50 × the **world pole vault record** (2014) (6.16 m)

The diameter of the **Death Star** in *Star Wars* (120 km) is
  100 thousand × the length of a **lightsaber**™ in *Star Wars* (1.19 m)

The Men's Olympic **long jump** record (8.9 m) is
  2.5 × the length of a standard full size **snooker table** (3.57 m)

The length of the **M25 motorway** around London (188 km) is
  400 × as long as the longest **golf drive** ever (471 m)

The Men's Olympic **triple jump** record (18.09 m) is
  10 × the width of **an ice-hockey goalmouth** (1.8 m)

The length of the longest **ski jump** ever (2015) (251.5 m) is
  5 × the typical flight of a **flying fish** (50 m)

The height of the **Empire State Building** (NYC) (381 m) is
  5000 × the size of a standard square **Post-It note** (76.2 mm)

The Third Technique: Divide and Conquer

Take One Bite at a Time

The Greek Theatre in the town of Taormina in Sicily has a spectacular view of the bay and of Europe's highest volcano, Etna.[28] The theatre itself dates to the third century BCE and, though called the 'Greek' Theatre, is largely the work of the Romans (the giveaway is that it is predominantly brick-built). It's regularly used as a concert venue, and the promotional material suggests that it originally had a capacity of 5000. Is this claim credible? Let's do a **cross-comparison** to see if we can make an independent estimate of our own to compare.

As the image shows, the seating is arranged in sections, seven of them in total. So, trying to arrive at our own estimate of the capacity, we can tackle the

---

[28] The height of Etna is 3350 m, more than a third the height of Mount Everest and rising straight from the sea.

simpler task of forming an idea of the capacity of one of those sections (and then later multiply by seven). Let's take a closer look at:

Counting the rows of seats (the lower ones are original stone, the upper ones are wooden bleachers), we get to 26 rows of seats currently in place. But there is some evidence of further, unrestored structure lower down—so we can guess that there might have been a further block of perhaps 12 rows there, for a total of 38 rows.

How many people might sit on one of those rows? More on the back rows, fewer in the front, but a reasonable figure for one of the middle rows might be 15 people.

So, we have 7 sections of 38 rows, each accommodating (on average) 15 people. Multiply those together to get 3990. It's the right order of magnitude, but somewhat short of the 5000 claimed: perhaps that's an over-optimistic claim?

But hold on! Those seven sections don't make a complete semicircle. There is, in fact, on each side, space for a further section, which would bring the total to nine sections,[29] and that would bring our estimate of the total number of seats to 5130. We can probably conclude that the claim of a capacity of 5000 is a fair one.

---

[29] Or perhaps those 'restricted view' areas would be standing room only. That would still count as part of the capacity of the theatre.

Let's reflect briefly on that approach: faced with a big number, 5000, a number that is a little difficult to swallow in one gulp, we set about breaking it into bite-sized chunks. The layout of the theatre is a gift as it allows us to reduce the problem by a factor of seven (or nine!) immediately. At that point, we are implicitly asking if each of those sections could accommodate 700+ people. Counting or estimating the rows allows us to shrink the problem even smaller, in effect turning it into an eyeball estimate of how many people could sit on an average-length row.

5000 is still a big number. But think of it as, more or less, 9 sections of 38 rows averaging 15 people each. None of those is a big number. We can easily find headroom for those.

Distributing the load

When shipping containers are stacked up to 12 layers deep on a container ship, it's important to keep the load balanced. Similarly, the fireman's lift is all about being able to carry a greater weight if the burden is properly balanced and centred. It's much the same with the divide-and-conquer technique for big numbers. 5000 theatre-goers is a big number, and it's all weighing on a single point, mentally. Redistribute that load as 9 × 38 × 15 and it becomes much easier to lift.

When we think about the world's population increasing from around 3 billion in 1960 to over 7 billion at present, we can use this trick: the 'billion' is carried on one shoulder, the 3 and the 7 on the other. When we decide to deal with land areas of countries in square kilometres rather than hectares or square metres, we are redistributing the load, making a big number smaller, at the expense of having also to mentally remember that we are working with a larger unit.

When the scientist expresses the speed of light as 1.08×10^9 km/h, she is dividing and conquering. The 'significand' of 1.08 on one hand, the 'power' of 9 on the other. One is used for close-call comparisons, the other for order-of-magnitude comparisons. We'll see this pattern over and over again as a way of comprehending big numbers. The point weight of the big number is replaced by a balanced load, and that increases our capacity for handling big numbers.

Pick your units and your multiples

One straightforward way of spreading the conceptual burden of big numbers is to pick your units to suit your task. Don't measure the area of a football field in square centimetres; don't measure the area of countries in hectares. One of the main aims of the International System of units (SI) is to make sure that by using the metric prefixes, you can (almost) always select a suitable unit. So we can, in theory, progress, in multiples of a thousand, from grams to kilograms to megagrams to gigagrams. This is an excellent scheme, but we've not really embraced it in everyday life—we substitute 'ton' for megagram, and we ignore bigger measures entirely.

The advice still holds. Choose your units wisely to allow yourself to work with numbers that are modest in size, balanced by your ability to visualise the unit you're working with. In this way, you can distribute the load of a big number into three parts: a significand that takes a value between 1 and 1000, a multiple (thousand, million, billion, etc.), and the biggest unit you're comfortable working with.

So this is the third technique: find ways to break the big number into parts, and get comfortable with the parts. Where numbers relate to areas or volumes, think about how big those shapes would be, in terms of height, width, and depth. Think about what units work best.

And, if you like, create some new units to use, as we did in the section on **visualisation** when we invented the ant, the beetle, the park, and the Oz, so that they become **landmark numbers**.

Ticking Away

How We Measure the Fourth Dimension

Which of these is the longest period of time?

☐ Time since the emergence of flowering plants
☐ Time since the earliest primates
☐ Time since the extinction of dinosaurs
☐ Age of earliest mammoth fossils

It's about time

How did it get so late so soon?   **unknown**

— There's a bristlecone pine tree in California that is thought to be almost 5000 years old. *Is that a big number?*

— Our Solar System orbits the centre of the Milky Way galaxy once every 240 million years. *Is that a big number?*

— The universe came into being 13.8 billion years ago. *Is that a big number?*

The wheels of time

In 1901, off the coast of the Greek island of Antikythera, a shipwreck of an ancient Greek ship was discovered, full of treasures. One unremarkable-looking piece—corroded brass and rotting wood—was ignored for 50 years, but it turned out to be the greatest treasure of all. Now known as the Antikythera

Mechanism, it is an astonishing piece of ancient engineering. It consists of a series of interconnected bronze cogwheels housed in a wooden box, with a crank handle, and dials on front and back, and is believed to have been constructed in the second century BCE in Syracuse in Sicily.

Analysis of the cogwheels and the inscriptions reveals that this is a device for calculating the position of the Sun and Moon, as well as making other calendar calculations such as the years for Olympic games,[30] and even eclipses. It is a truly remarkable object, and nothing comparable was made until well over a thousand years later. As Syracuse is known to be the home of the mathematician Archimedes, and the mechanism has been dated to the century after his death, it's not too fanciful to imagine that it might be based on his work.

The mechanism is a device for visualising the passage of time by simulating the movements of the heavenly bodies that are so important in the regulating the way we measure time, and order our earthly affairs. It's clearly an item of great prestige, and high craftsmanship, and shows how important it was to keep track of time, and its complexities, even 2000 years ago.

Time is change

Time is change; we measure its passing by how much things alter. **Nadine Gordimer**

Relativity teaches us that time and space cannot be disentangled, that time is one dimension of a unitary space–time, but it's not that simple. We can't, in practice, measure time as we can the three spatial dimensions. We can't take a measuring rod and count off a number of units. The only way to measure time is to wait for it to pass. And once it has passed, we can never revisit it. So we can never repeat measurements of time.

We know time passes, because we are aware of change. Measuring time is measuring change—either change as shown by a process that we believe to be repeating in a regular way, or change manifested in a process that progresses at what we believe to be a constant rate. Our timepieces rely on mechanisms or processes that are periodic (such as a pendulum ticking away the seconds),

---

[30] Every year, the Greeks held a festival of games in one of four locations. So the Olympic games were part of a four-year cycle, as they are now. The mechanism has a display disc sectioned into four quadrants, labelled NEMEA, ISTHMIA, PYTHIA and OLYMPIA.

in which case we can count the periods, or processes that progress at a steady pace (a candle burning the midnight hours away), in which case we can measure the progress by measuring linear distance (the remaining height of the candle).

Luckily, the natural world changes constantly and is full of events that progress steadily and that repeat regularly. Every day, the Sun brings light into the world, and every night the light leaves. Twice a day, the tides surge in and retreat. Every month (more or less), the Moon grows and shrinks. Every year is marked by the progressing of the seasons. Every second or so, my heart thumps.

Unfortunately, none of these cycles offers a particularly convenient basis for a coherent system of measurement. The lunar month isn't a whole number of days, nor are our bodged calendar months of equal length. The lunar months don't subdivide the year properly. The tides don't partition the day evenly, nor does the Sun rise at the same time every day.

But timekeeping has always been important. Some of the earliest scientific and mathematical studies were concerned with mastering timekeeping. The great Islamic mathematician, astronomer, and poet Omar Khayyam was part of a committee who worked on calendar reform in 1079 in the imperial observatory in Isfahan in what was then Persia. The outcome was the Jalali calendar, one of the most accurate solar calendars ever devised. The same committee also calculated the length of the year to an accuracy of about 1 part in 10 million.

Giving the time of day

For all the complications involved in tracking the heavens, one specific kind of measurement proved sufficiently reliable and regular: observing the moment the Sun reached its maximum height on any day—at noon. Regardless of the time of year, the period from one noon to the next stays more or less the same,[31] and that makes the 'solar day' a good starting point for a system of time.

---

[31] In fact, this is far from exact. If you plot the position of the Sun in the sky at noon each day over the course of a year, it describes an elongated figure-of-eight shape called an analemma. At worst, this puts the Sun at noon 4° East or West of true South (or true North in the Southern Hemisphere), which corresponds to a sundial inaccuracy of around 15 minutes.

But how to subdivide the day? Some of the earliest schemes, for example that used in Ancient Egypt, involved dividing hours of daylight into 12 equal parts, and the hours of night into 12 parts too. This meant that for most of the year, the night-time hours were of a different length to the daytime ones.[32] In time, this system was reformed and an 'equinoctial' hour was adopted, which was based on the uniform length of the hours observed on the days of the spring and autumn equinox. Twelve is a convenient number to use for division of the day (and night) into shifts and watches,[33] as it can be evenly divided into two, three, four or six parts. This eventually led to our current division of the day into 24 equal hours.

Before mechanical clocks, timekeeping needed some mechanism or process that reflected slow and steady change. The most obvious and direct of the measuring tools was the sundial, which simply measures the progress of the Sun across the sky, as long as the Sun is visible.[34] Other time-telling mechanisms were a variety of candle clocks, water clocks, sand-timers, and so on. All of these were analogue processes that converted the time elapsed into a measure of linear distance.

Any inaccuracy in these very approximate devices could naturally be re-set by observing the Sun at its highest point when noon came around. However approximate your time-teller, there would always be an opportunity for recalibration.

During the Middle Ages, people started writing about a smaller division of the hour, the *pars minuta prima*, or the 'first small part', as one-sixtieth ($^1/_{60}$) of an hour, which of course became our minute. The 'second small part' became our 'second'.

So the second, as a unit of time, while fundamental to our modern timekeeping systems, seems to have no inherent connection to any everyday phenomenon. However, it sits well as a human-scale measure. It is approximately the right size of unit to use for human counting, and it's close in size to the period of the

---

[32] The Japanese had a similar arrangement: the British Museum houses a Japanese clock dating from the eighteenth century, which has an exposed mechanism to allow twice-daily adjustment of its speed. So it can measure six periods from sunset to sunrise, and six periods of different length from sunrise to sunset, regardless of the time of year.

[33] The word 'watch' shares an etymological connection with 'wake'. The use of this word to mean a small timepiece seems a natural extension of meaning, but the exact route by which this happened is uncertain.

[34] One of the reasons sundials are at all effective is that they are, at least approximately, independent of the seasons. Even though the height of the Sun at a given time of day varies dramatically through the year, the (horizontal) bearing of its position at each hour varies only by a few degrees either way.

human heartbeat and breath. This human scale and utility validated its eventual acceptance as a base unit.

If sundials, water clocks, and so on are essentially analogue devices, all the 'ticking' mechanical clocks are essentially digital even if their displays are not. How so? Well, they all involve a periodically repeating component—for example, a pendulum—and this periodic movement is coupled to an 'escapement' mechanism (the ticking) that digitises continuous time by chopping it into discrete chunks, and this ticking drives a counting device. The count, always a whole number of ticks, could be (but was not always) displayed using rotating hands, and the counter could also trigger events such as the ringing of bells.[35] The first such mechanical clock was made in China in 725 CE, and is credited to Yi Xing, a renowned scholar. The cogwheel clock was powered by hydraulics, and to overcome problems with freezing, a model using mercury instead of water was made for winter use. The earliest clocks powered by falling weights were made around 1300 CE.

Landmark number First mechanical clock—725 CE

Although the second is included as a base unit in the International System of units (SI), it doesn't follow the standard rules of tenfold multiples that govern the rest of the SI. This is the one area where traditional measures remain unconquered by the metric system. Where are the kiloseconds and the megaseconds?

It's not that the great reformers of measurement in the wake of the French Revolution didn't try to bring time into the great decimalisation project. In 1793, a decree was issued in France that established the decimal hour (one-tenth of a day), the decimal minute (one-hundredth of a decimal hour), and the decimal second (one-hundredth of a decimal minute and equal to 0.864 of our seconds). Decimal time was only officially used in France for a little over six months, from September 1794 until April 1795. However, at least one person got the habit. It is known that the mathematician/astronomer Laplace, whose career in France spanned periods before and after the revolution, used a decimal watch and recorded his astronomical observations using fractional days, in effect using a decimal time system.

The French reformers weren't the last to try this decimalisation of time. In 1998, in the heady days of the first Internet boom, the Swiss watch company Swatch launched a range of watches under the label *.beats*. These watches didn't

---

[35] The word clock comes from the Latin *clocca*, a bell.

use boring old hours, minutes, and seconds to tell time. They used '.beats'. One '.beat' was a thousandth part of a day, so 86.4 seconds long. Times would be written with the oh-so-cool '@' symbol, so that noon would be @500. Not only that, there would be no time zones (the favoured meridian would be that used by Swatch headquarters, UTC+1) so that the time in '.beats' would be the same all around the world. Unsurprisingly, this so-called Internet time failed to catch on, although you can still find it as a curiosity in some corners of the web.

But decimal time is probably still part of your life. Whenever you enter a time into an Microsoft Excel spreadsheet, for example, it is stored as a fraction of a day. Try entering 12 noon ('12:00') into a cell of a spreadsheet (say into cell A1) and then, in a different cell, enter a calculation such as '=2.5 * A1'. You'll see it displayed as '06:00'. What's happened is that the spreadsheet has multiplied 0.5 (12 noon) by 2.5 to get 1.25. It has then displayed only the fractional part, 0.25, which it interprets as 6 a.m. (a quarter of the way through a day).

> **Landmark numbers**
> - There are 1440 minutes and 86,400 seconds in a day.
> - There are around half a million (525,600) minutes in a 365-day year.
> - There are 31.536 million seconds in a 365-day year.

It's that time of year

So, the second as a unit of time is arbitrary in size—convenient but arbitrary. However, the two measures of time that are truly fundamental to life on Earth are anything but arbitrary—the length of a day and the length of a year. It really is extremely inconvenient that these two measures are not commensurate with one another. Nonetheless, the inherent benefit and logic of using the day and the year as our units of time is so compelling that we go to great lengths to make them fit together. This has led to numerous calendar revisions and adjustments, for example the introduction of leap days and leap seconds, all to retain the benefits of keeping our time systems properly aligned to the inflexible days and years. There really is no alternative, and so we live each day at a pace dictated by the rotation of the Earth about its axis, and every year to a cycle dictated by the journey of the Earth around the Sun.

Only slightly less important in our measurement of time is the cycle of that other dominant presence in Earth's sky, the Moon, visibly going through its wax-and-wane cycle every 29.53 days, and governing the near-monthly cycle of the tides.

Landmark number Length of a lunar month—29.53 days

In the fifth century BCE, Babylonian astronomers tracked the course of heavenly bodies through the year by reference to the band of stars visible in the plane of rotation of the Earth around the Sun, the ecliptic, which served as fixed reference points. They divided this circle into 12 segments of 30° each, and named each segment for a constellation of stars found in it. When the Greeks encountered this system, they called it the *zōidiakòs kýklos*, ('circle of little animals'), and we now know it as the Zodiac, and (regrettably) some people still make use of it in popular astrology. How much ingenuity and arithmetic ability has been wasted on such ultimately pointless calculation?

Shakespeare was right, of course, when he had Cassius say 'the fault...is not in our stars, but in ourselves'. In our lives, the stars thankfully do not govern us, but, when it comes to measuring time, we really have been greatly influenced by the heavenly bodies.

Some things just won't fit

The persistent problem is that the Earth's spin cycle (day), the Moon's phase cycle (month), and the Earth's orbital cycle (year), are not commensurable—and nothing will ever make them so. This has given a hard job to the makers of calendars through the ages. How do they decide what whole number of days to fit into a month (and not displease the Moon too much), and how to make a whole number of months fit into the year (and not allow the seasons to drift)?

The Romans ascribed the first of their calendars to Romulus, the legendary co-founder of Rome. In his scheme, March was the first month, and there were only ten named months, totalling 304 days, with the remaining 61 'winter days' allocated to no month at all. The months after June were named for numbers: what we now would think of as July was Quintilis (fifth) and of course the tenth month was December. Before very long, the second of Rome's kings, Numa Pompilius, (reputedly) fashioned the winter months into two new months at the start of the year, naming them January for Janus, the doorkeeper of the gods, and February for Februa, a festival of purification. King Numa's calendar consisted of 12 months: seven with 29 days, four with 31 days, and February with 28 days. That, you may calculate, came to 355 days, a full ten days (and a bit) short of the actual year. To correct this, in some years an intercalary month was added, after February. This month, Mercedonius ('work month') was sometimes manipulated as a political device whereby the *Pontifex Maximus* (chief priest) could extend or cut short terms of political office at will. This confused situation—it became so bad that citizens outside Rome were sometimes left ignorant of the 'true' Roman date—was improved hugely by the reforms introduced by the energetic Julius Caesar when he served as *Pontifex Maximus* in 46 BCE.

The Julian calendar dealt with the worst of these problems—no extra months needed, just a leap day every four years—and prevailed in Europe until 1582 when reforms under the authority of Pope Gregory XIII gave us the Gregorian calendar.[36]

This Gregorian legacy has become our conventional Western calendar. We still have the Romans to thank for the eccentric pattern of the days in the months, with that infuriatingly non-memorable mnemonic rhyme ('Thirty days hath September...'), and the fudge that is the leap year.

We can't blame the Romans for the complexities of calculating Easter in the Christian calendar. That's all to do with the interaction between the lunar month and the solar year. It affects the Christian church calendar, the Islamic calendar and the Jewish calendar in different ways.

The Islamic calendar is primarily a lunar calendar, with the months defined by the cycles of the Moon. Twelve months add up to a year, but twelve lunar months are shorter than a solar year. So the months of the calendar move through the seasons in a cycle that repeats every 33 years. Notably, this means that the month of Ramadan, when devout Muslims fast from sunrise to sunset, can occur both in the winter, with short days (making for an easier fast) and in summer with its longer days (giving a more onerous fast). It also means that the Islamic calendar's count of years progresses more rapidly than the conventional Western calendar: 33 years in the Gregorian calendar will equate to 408 lunar months and so 34 years on the Islamic calendar.

One of the output dials on the Antikythera Mechanism shows the progression of what's known as the Metonic cycle. It's long been known that lunar months don't fit evenly into the (solar) year, but Meton of Athens, working in the fifth century BCE, found a very useful way of making adjustments to allow for this. He identified that 19 solar years correspond almost exactly to 235 lunar months. His calculations meant that to keep a lunar calendar on track, you need to add an extra lunar 'leap' month in 7 out of 19 years.

---

[36] Under the Julian calendar, every fourth year was a leap year. The Gregorian calendar removed three of these leap years in every 400, the ones that fall on years divisible by 100, but not 400. Thus 1700, 1800, and 1900 were not leap years (and nor will 2100 be), but 1600 and 2000 were.

The Jewish calendar, like the Babylonian calendar on which it is modelled, is also a lunar calendar, but in this case to stop the drift of the months through the seasons of the year—custom says that Passover must (in the Northern Hemisphere) remain a spring festival—the calendar uses a trick based on the Metonic cycle: in 7 out of every 19 years (specifically years 3, 6, 8, 11, 14, 17, and 19) a 13th intercalary month is inserted to pull things back into alignment. This is the same cycle that is used for the calculation of Easter in the Christian calendar, which explains why the date of Easter skips about in the calendar so much.

> **Landmark number** 19 solar years is 235 lunar months: the Metonic cycle.

Turning years into numbers

History is measured in years. The human race has been writing things down for around 5000 years, which is a big number, but not very big. For the purposes of history, societies around the world have tended to use some fixed date in the past as the 'zero' point for our counting of years.

The year numbering system that most of us use is based originally on the supposed date of the birth of Jesus of Nazareth. The designations BC (before Christ) and AD (*anno domini*, in the year of our lord) are now more generally superseded by BCE (before the common era) and CE (in the common era).

The Romans counted years AUC (*ab urbe condita*, from the founding of the city), which event was deemed to have happened in what we would now call 753 BCE; the Muslim calendar counts from when Muhammad moved from Mecca to Medina, in 622 CE; the Jewish calendar is based on a deemed date of creation, reckoning backwards from biblical authority: this works out to 3761 BCE.

In China, the ancient and historical system worked by reference to eras based on the year of accession of the then-current emperor. The last such date was 1908, the start of the short-lived Xuantong Era under the 'Last Emperor'. In 1912, the first president of the Republic of China, Sun Yat-Sen, rebased the year number to 2698 BCE. This corresponds approximately to the supposed start of the reign of the legendary Yellow Emperor.

If the Ancient Mayans have been in the news recently, it's for one reason: their calendar and the prognostications of doom that would supposedly befall the world when their Long Count calendar 'clocked over' on 21 December

2012. In fact, this was simply the equivalent of the most significant digit of your car's odometer moving to the next digit (in this case, a 12 became a 13 in a base-20 counter). The Long Count as a whole will not clock over fully until the year 4772 CE,[37] having notionally started in 3114 BCE.

Measuring prehistory

When we look to times long before recorded history, to prehistoric and geological time, the particular choice of a zero point for the calendar becomes unimportant. When discussing things that happened a million years ago, the odd few thousand here and there makes little difference. Instead, things are generally dated by years before the current date. Since the year is not an SI unit, its use as a unit is not regularised by that system: there are multiple systems and conventions in place for denoting 'years before present'. Although there is an ISO standard validating the 'annum' (symbol 'a') as equal to one year, you might still find all of the following for denoting the time of the end of the Cretaceous period:

- 66 million years ago
- 66 mya
- 66 Ma ago (mega-annums)
- 66 million BP (before present)

Other than the thousand, million, and billion multiples, the year is as large a measure as modern science generally adopts when it comes to time. True, there is a 'galactic year', being the time that our Sun (and its attendant system) takes to make an orbit of the galactic centre, and equal to 225 million years, but this is not used as a unit in its own right.

> **Landmark number** Galactic year—225 million years.
> The Earth is now around 20 galactic years old.

In charting the history of the Earth, though, the geologists have devised a hierarchy of ever-larger time classifications to frame their analysis of the major changes that the Earth has undergone since it was formed. The smallest 'unit' in

---

[37] Nor should this date be a cause for alarm. The Mayans had four more levels of counting cycles in addition beyond this!

this system here is the 'Age' (although these are not really units as there is no consistency as to length—it's simply a hierarchical grouping):

- Ages are collected into 'Epochs'
- Epochs into 'Periods'
- Periods into 'Eras'
- Eras into 'Eons'
- The first eon is termed a 'Supereon', the Precambrian

Time and technology

Astronomy and timekeeping were among the principal drivers for the development of arithmetic and mathematics in ancient times. Technology and timekeeping have always gone hand in hand, as the Antikythera Mechanism shows. The primary function of clocks has always been simply to keep counting, and to allow us to read off precisely how many seconds have passed. When the first electronic circuits were developed, it was inevitable that timing circuitry would be needed. And so time became fully digital, and a clock became no more than an electronic circuit counting the elapsed seconds from some zero point.

Most modern computer systems now represent time at the lowest level by reference to the number of seconds since an 'epoch date', a selected zero point. Many computers in use today are built on some version of the operating system Unix and make use of 'Unix time' to keep track of dates and times. Unix time counts the seconds since 1 January 1970.

Date and times before that time are represented as negative numbers, but these times extend back only as far as 13 December 1901. Fractions of seconds are represented as counts of millionths or billionths of seconds.

Computers store numbers in 'registers', which have a maximum length. Where the Unix time is stored in a 32-bit register, which is very common, this representation will fill up the register and reach its maximum on 19 January 2038, when the number will, unless prevented from doing so, 'roll over' and start again from 13 December 1901.

This is a problem very similar to the 'Millennium Bug'. That was a situation that developed in the late twentieth century. In a time of severely restricted computer memory and bandwidth, every effort was made to save storage, even to the point of storing year numbers as two digits. After all, if you were writing a computer program in the year 1984, say, it was unimaginable to most programmers that their code would still be in use 16 years later. But very many so-called legacy systems were indeed in place in 1999, and the danger was that as two digit years rolled over from '99' for '1999' to '00' for '2000', all sorts of date and age calculations would fail. At enormous cost, many thousands of systems were adapted and the crisis was averted. Will a similar problem arise in 2038? It's possible.

The far future

Time had a beginning. Cosmology tells that the absolute zero of time, the Big Bang, was 13.8 billion years ago. This makes the age of the universe around 4.5×10^{17} seconds—less than a quintillion seconds.

Douglas Hofstadter, in his book *Metamagical Themas*, illustrates the difference between millions and billions and why we sometimes disregard that difference:

The renowned cosmologist Professor Bignumska, lecturing on the future of the universe, had just stated that in about a billion years, according to her calculations, the Earth would fall into the sun in a fiery death. In the back of the auditorium a tremulous voice piped up: 'Excuse me, Professor', but h-h-how long did you say it would be?'' Professor Bignumska calmly replied, 'About a billion years.' A sigh of relief was heard. 'Whew! for a minute there, I thought you'd said a million years.'

In fact, Professor Bignumska is a little pessimistic. Current thinking is that we can expect that our Sun will last for a further five billion years or so. In its death throes, the Earth will be made uninhabitable. We'd better get going to ensure our survival beyond that!

And after that? Will time end? Does time have a biggest number? The truth is: we don't know. It's one of the great unanswered questions. Physicists are currently wrestling with the evidence that there is both more matter in the universe than is visible ('dark matter') and more energy than can be accounted for ('dark energy'). Understanding these topics may provide some understanding of how the universe will develop from here, and whether the universe's lifetime will be measured in billions or trillions of years, or numbers even bigger than these.

Landmark number
Our Sun is probably around half-way through its lifecycle. Very roughly, 5 billion years gone, 5 billion years to go.

What happened a thousand years ago?
A number ladder for time

100 y	Time since first fixed-wing scheduled air service (102 years)
200 y	Time since first iron steamship crossed the English Channel (194 years)
500 y	Time since birth of Copernicus (543 years)
1000 y	Time since start of building of Great Zimbabwe (1000 years)
2000 y	Time since start of building of the Colosseum (1944 years)
5000 y	Time since building of the Great Pyramid of Giza/Cheops/Khufu (4580 years) Time since building of Stonehenge (5120 years)
10,000 y	Time since earliest farming (11,500 years)
20,000 y	Time since first humans in the Americas (15,000 years)
50,000 y	Time since first humans in Australia (46,000 years)
100,000 y	Start of most recent period of glaciation—'Ice Age' (110,000 years)
200,000 y	Time since first modern humans (200,000 years)
500,000 y	Age of earliest Neanderthal fossils (350,000 years)
1 million y	Earliest evidence of fire-making (1.5 million years)
2 million y	Time since first members of genus *Homo*—first humans (2.6 million years)
5 million y	Age of earliest mammoth fossils (4.8 million years)
10 million y	Human lineage diverged from chimpanzees (before 7 million years)
20 million y	Time since end of the Paleogene geologic period (23 million years)
50 million y	Time since earliest primates (75 million years)
100 million y	Time since emergence of flowering plants (125 million years)
200 million y	Break-up of Pangaea into today's continents (175 million years)
500 million y	Time since emergence of earliest fish (530 million years) Explosion of marine algae (650 million years)
1 billion y	Divergence of eukaryotes into ancestors of plants, fungi, and animals (1.5 billion years)

2 billion y   Time since emergence of multicellular life (2.1 billion years)

5 billion y   Time since start of formation of Solar System
(4.6 billion years)

10 billion y   Age of universe (13.82 billion years)

Goodness, Is that the time?

Time since first **humans in the Americas** (15,000 years) is
500 × **lifespan of a horse** (30 years)

Time since **earliest writing** (5200 y) is
25 × time since birth of **Darwin** (208 y)

Length of **reign of Britain's Queen Victoria** (63 years) is
2.5 × the **average human generation** (25 years)

Time since start of **building of the Colosseum** (1944 years) is
10 × time since **first iron steamship** crossed the English Channel (194 years)

Time since birth of **Diophantus**[38] (1810 years) is
4 × time since birth of **Shakespeare** (452 years)

Period of **Saturn's orbit** (29.5 years) is
2.5 × period of **Jupiter's orbit** (11.86 years)

Time since birth of **Archimedes** (2300 years) is
4 × time since invention of **printing press** (576 years)

Time since invention of **printing press** (576 years) is
5 × time since first transatlantic **wireless transmission** (115 years)

---

[38] Now go and Google Diophantus.

An Even Briefer History of Time

Some of the big numbers that we find hardest to grasp are those to do with history (especially ancient history) and prehistory. From the perspectives of human lifetimes, it's hard to appreciate the immense stretches of time that led to us being here, now.

Here then are some numbers to help you find your way when you go time-travelling. A disclaimer: it's a nice paradox that as we make our way into the future, we learn more and more about the past. So the timelines set out here reflect the information available at the time of writing—undoubtedly discoveries will still be made that will shift some of these numbers around. But even if the fine detail will change, it's still worth painting the big picture, so far as we know it. So we'll find a few **landmark numbers**, and do a little **dividing and conquering.** Some **visualisation** might help, too, in trying to tame the big numbers of the deep past.

Geological time by the numbers

For all that we've accomplished, humans have walked the Earth for only a minuscule fraction of its existence. Paleontological and archaeological discoveries, however fascinating, can be difficult for non-experts to put into context. So here are a few pages of essential **landmarks** and diagrams to help you **visualise** the timeline of our planet, and how life emerged, from the Big Bang up to the Stone Age:

- The universe came into being 13.8 billion years ago.
- The Earth formed 4.57 billion years ago. This means the Earth has been around for about a third of the universe's existence.

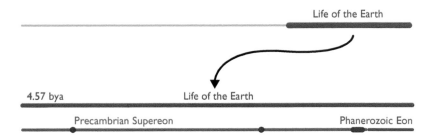

- The first 4 billion years (88%) of the Earth's existence is called the Precambrian Supereon,[39] and lasted until 541 million years ago. That's eight-ninths ($^{8}/_{9}$) of the time since the Earth was formed. Highlights of the Precambrian:
  - First **simple single-celled** life (bacteria)—4 billion years ago
  - First **complex single-celled** life—1.8 billion years ago
  - A hypothesised 85-million-year period nicknamed 'Snowball Earth'[40] around 700 million years ago
  - First **multicellular** life—635 million years ago

Landmark numbers
- First life—4 billion years ago
- First multicellular life—635 million years ago

- Unsurprisingly, the period immediately following the Precambrian is the Cambrian, the first period of the Phanerozoic Eon.[41] The Phanerozoic

---

[39] A supereon is, well, just a larger unit of organisation than an eon.
[40] This refers to a period when the Earth's surface is thought to have been entirely frozen. It's speculated that the freezing conditions may have contributed to the evolution of multicellular life.
[41] Eons are divided into eras, which are divided into periods, which are divided into epochs.

Eon (the name means 'visible life')[42] is the eon we are in currently, and it has so far lasted 541 million years, which is nonetheless less than ⅛ of the whole span of the Earth's existence.

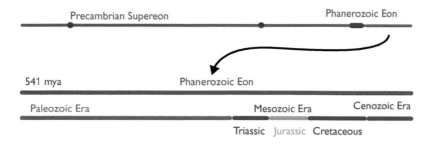

- The first era in this eon has been named the Paleozoic,[43] and it lasted 289 million years, over 6% of the Earth's lifetime. The starting point of this Paleozoic Era is the huge diversification of life forms known as the Cambrian explosion and evolution of a huge variety of life forms (but not yet any dinosaurs).

> **Landmark number** The Paleozoic Era ended 252 million years ago with a massive extinction, the Permian–Triassic extinction event. The causes are unknown, but 96% of marine species died off, as well as huge numbers of land species. Recovery took millions of years.

- After the Paleozoic came the Mesozoic[44] Era, which lasted 186 million years, 4% of the Earth's lifetime. The Mesozoic is divided into three periods. Going forward through time:
  - The Triassic: Early dinosaurs, pterosaurs, first mammals. Super-continent of Pangaea breaks into Laurasia and Gondwanaland.
  - The Jurassic: All your favourite dinosaurs, early birds. Map of the Earth as we know it is not yet recognisable.
  - The Cretaceous: Fish, sharks, crocodiles, dinosaurs, birds prevalent. Map of the Earth is recognisable, but distorted.

---

[42] These seemingly obscure geological names, when decoded from their Greek roots, hold useful clues. In this case, Phanerozoic = 'visible life' marks the first macroscopic organisms.
[43] Paleozoic = 'old life'.
[44] Mesozoic = 'middle life'.

> **Landmark number** The Mesozoic Era was the age of the dinosaurs, and its end is marked by their extinction—66 million years ago. After this, mammals become prevalent.

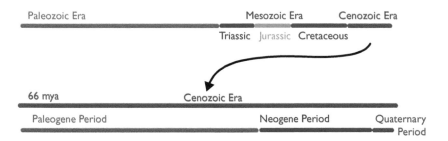

- We now reach the third era of the Phanerozoic Eon, the Cenozoic[45] Era, which is still ongoing. The first two periods in this era are named the Paleogene (which lasted 42 million years) and the Neogene (20 million years). Through these periods, mammals have thrived, and the world has become populated with animals broadly familiar to us. So far, the Cenozoic Era has lasted 66 million years, just 1.45% of the Earth's lifetime.

> **Landmark number** On land at least, this era, starting 66 million years ago, is the time of the mammals.

- Eras are divided into periods, and we're now living in the Quaternary[46] Period, which started 2.6 million years ago. The Quaternary is split into the Pleistocene[47] and Holocene epochs. The Pleistocene includes the time known as the Ice Age,[48] and saw the emergence of modern humans. It lasted 2.58 million years, almost the whole of the Quaternary Period. The current epoch, the Holocene, is a small fraction of that.

---

[45] Cenozoic = 'new life'.

[46] The name is a holdover from an earlier naming scheme. It is no longer the fourth of anything.

[47] Pleistocene = 'most new'.

[48] The popular term 'Ice Age' refers to the most recent period of glaciation, lasting from approximately 110,000 years ago to 11,700 years ago.

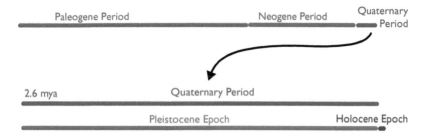

- We're now living in the Holocene[49] Epoch of the Quaternary Period. This epoch started after the end of the last major glaciation ('Ice Age'). During this epoch, humans have moved out of the Stone Age, and so far this epoch has lasted 11,700 years.

It's been proposed (but is not yet part of official nomenclature) that a new epoch be designated, the Anthropocene, reflecting the impact that humanity now has on the planet. The term is now in widespread use informally, but as yet has no formal definition or agreed start date.

Human prehistory by the numbers

The evolution of *Homo sapiens*

Picture a space rocket: the *Saturn V* that sent three clever apes to the Moon in 1969 will do nicely. That rocket comprised multiple stages. The first stage was massive, essentially nothing more than a huge fuel tank with a burner at the back. The second stage was smaller, the third stage smaller still, and only then do we come to the payload with the tiny capsule at the tip, home for a few days to the three travellers. All that mass, that bulk, was merely the necessary preamble to the human story of that adventure. The story of human evolution to date feels a bit like that: an eternity of preamble, that led eventually to the creatures we would call modern humans.

The rocks have told us the story of how it took billions of years for complex, macroscopic life to develop. It took hundreds of millions of years more to reach the emergence of the first primates. From those primates to the creatures whose

---

[49] Derivation from the Greek: Holocene = 'entirely new'.

evolutionary path forked away from the chimps to become *Homo* took millions more years. In our story, those booster stages have now done their work, and we are getting to the pointy end of the rocket.[50] This is the start of the human story.

The Neogene period started 23 million years ago (that's almost precisely $^1/_{200}$ of the Earth's age). At the start of that period, the great apes had not yet differentiated from the monkeys (something that happened only around 15 million years ago). But during the Neogene, and by 7 million years ago, our lineage became distinct from the chimpanzees.

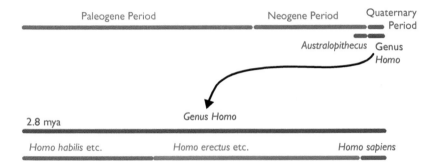

- Around four million years ago,[51] a genus of apes that we call *Australopithecus* evolved in East Africa. They walked on two legs, and are thought to be the ancestor genus to humans.
- The first creatures we call humans (genus *Homo*) are dated to around 2.8 million years ago. This means that for 99.94% of this planet's existence, there were no humans on it. The species we call *Homo habilis* (handy human) was making stone tools in the late Neogene Period, but, roughly speaking, the start of the human genus corresponds with the start of the Quaternary Period (2.6 million years ago). This was the start of the Paleolithic[52] Age, the Old Stone Age.

> **Landmark number** Earliest humans, start of the Quaternary Period, start of the Paleolithic Age—3 million years ago, very approximately

---

[50] Of course, unlike NASA engineers, evolution is not purposeful. The past history of our species has shaped us, but not intentionally.

[51] That's about one-thousandth of the age of the Earth.

[52] Paleolithic = 'old stone'.

- *Homo erectus* (upright-walking human) is an umbrella term for several species that evolved around 1.8 million years ago in the Quaternary Period. Around 1.5 million years ago, there is the first evidence for firemaking.
- Early forms of humans were spreading from Africa to Eurasia. There is evidence of humans in Europe 600,000 years ago, likely a common ancestor to both modern humans and Neanderthals.
- The earliest evidence of our own species, *Homo sapiens* ('wise human'), dates to about 200,000 years ago in East Africa.[53] Although at this time *Homo sapiens* was already anatomically the same as a modern human, it would be another 150,000 years before there is evidence of what we'd call modern ways of thinking.

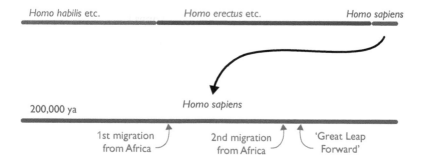

- Evidence is still unclear, but it seems that two waves of migration of modern humans from Africa occurred, the first about 120,000 years ago, the next around 60,000 years ago.
- It seems that at some time after these migrations, around 50,000 years ago, *Homo sapiens* acquired behaviours and cognitive capabilities that we would recognise as human, such as abstract thinking, planning, and art. The cause of this so-called great leap forward is unclear, but one possibility is the impact of cooking on quality of nutrition.

> **Landmark number** *Homo sapiens*, modern in mind and body, dates from 50,000 years ago.

[53] Breaking news (at the time of writing): in June 2017, fossils found in Morocco suggest a date as early as 300,000 years ago for the earliest *Homo sapiens*. The past is always changing!

The big migration

Humans evolved biologically over billions of years. But once humans and human communities had reached the level of behavioural modernity, roughly 50,000 years ago, it became possible for development to proceed through a kind of cultural evolution. The species developed and expanded in an unprecedented way.

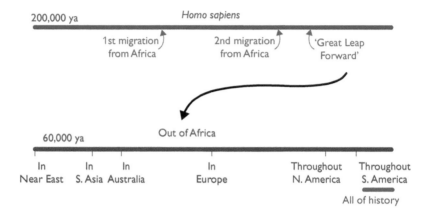

- Early humans (*Homo*, but not yet *sapiens*) spread from Africa around 1.5 million years ago and were the progenitors of populations in Europe and Asia.
- *Homo sapiens* arose in Africa and started to spread into the Near East 120,000 years ago, but this migration fizzled out.
- About 60,000 years ago a migration started, with humans travelling East from the Horn of Africa, into Yemen, and so onward to reach South Asia by 50,000 years ago.
- The first modern humans reached Australia by 46,000 years ago.
- The first *Homo sapiens* to reach Europe has been called Cro-Magnon[54] man, coming from the direction of modern Turkey. The Cro-Magnons came into a world where Neanderthals were already present, and by 30,000 years ago, they were widespread through Europe.

---

[54] The name comes from the location in southwestern France where the first specimens were found.

- Modern humans also spread north and eastwards, reaching Siberia and Japan by 35,000 years ago.
- At some point not later than 16,000 years ago, humans had crossed a land bridge from Siberia to Alaska, entering North America.
- Migrating south through America, humans populated the bulk of what is now the USA by 11,000 years ago.
- Humans had reached South America by 6000 years ago.
- Writing was invented around 5200 years ago, and history began. Historical times account for roughly 1 millionth part of the Earth's history.

Landmark numbers
- Modern humans in Near East: 60,000 years ago
- Modern humans in South Asia: 50,000 years ago
- Modern humans in Australia: 45,000 years ago
- Modern humans in Europe: 30,000 years ago
- Modern humans throughout North America: 11,000 years ago
- Modern humans throughout South America: 6000 years ago

Technological landmarks

- Earliest stone tool use—3 million years ago
- First fire-making—1.5 million years ago
- First sewing needle—50,000 years ago
- First ceramics—27,000 years ago
- Domestication of animals—17,000 years ago
- First copper-working—11,000 years ago
- First bow—11,000 years ago
- First agricultural revolution—10,000 years ago
- First numerical record-keeping—10,000 years ago
- First use of the wheel—6000 years ago
- First writing—5200 years ago, around 3200 BCE
- Bronze age—from around 3000 BCE until 1200 BCE
- Iron age—from around 1200 BCE
- Glass vessels—1500 BCE
- Weight-driven clocks—around 1200 CE
- Guns—around 1300 CE

Ancient history by the numbers

When looking at the past, we can choose from a range of lenses. A close-up view might show how life was for a particular group of people at a particular time in the past. A middle-distance view might analyse some specific episode or chain of events, such as the causes and consequences of the Boer War (1899–1902). Or we can use a wide-angle lens and take a panoramic view of humanity over time: the rise and fall of empires.

All three approaches have their merits, but, for this book, the big picture is what we're after. In building our numerate world-view, we need a framework for new information to find its place, or be challenged. We want the landmarks. We're not too fussed with specific dates ('more or less' is good enough), but what we're after is the broad sweep of history. So we'll take the panoramic view, and apologise in advance if the lens is too wide to capture much fine detail.

When looking into the past, at what point are we out of our depth? Where do the big numbers begin? I'll take the view that, as in other topics, the big numbers start at 1000. So this section will cover a period from the approximate start of writing (5000 years ago), and we'll take it forward to around 1000 years ago.

It's with some trepidation that I step onto the turf of 'history'. This is certainly not a history book. But this project is all about using numbers to provide context to understand the world we're in, and that has to include the big numbers that measure our past. To understand who we are and how we got to where we are, some historical context is essential. At the very least, this means putting past events in sequence, on a timeline. Or rather a series of timelines, since the human story is a braid of different narrative threads, sometimes separate, sometimes entwined.

This section sets out to show which cultures and civilisations were making their mark across the world in these successive 'time slices' of history: 3000–2000 BCE; 2000–1000 BCE; 1000–500 BCE; 500–1 BCE; 1–500 CE; and 500–1000 CE. This is necessarily an enormous oversimplification. I've crudely reduced history to a series of charts and bullet points. Any one of those bullet points stands for a distinct and fascinating conjunction of place, time, and culture, in which thousands of people lived over hundreds of years, and about which volumes could be and have been written. You may feel I've also done a disservice to the thousands of smaller cultures and peoples who did not fall within the compass of one of the larger empires or nations that appear in the lists, or who have not been studied and documented as thoroughly. And indeed I have.

Enough apologising! We're painting with a very broad brush here, and using just a few strokes. I hope, though, that this brief overview can create at least an impressionistic picture that gives an idea of how things fit together, of when and how the dominant cultures rose and fell through the four thousand years of ancient history.

5000 years and more ago (3000 BCE and before)

Nothing happened. To a good approximation, there is no written history earlier than 3000 BCE. This is not to say that people didn't live and love, fight and trade before then, but they didn't write it down. Writing was, as far as we know, invented around 3200 BCE. It's true that we know a considerable amount about earlier times: but this knowledge is indirect and derives from interpretation of the archaeological evidence, sometimes meagre, sometimes extensive. This has meant that we have had to deduce how people lived and developed through those times, rather than read about it in their own words.

But writing makes all the difference. Writing, for record-keeping, seems to have been an essential factor in the development of the apparatus of large states, and it's no coincidence that writing and the first empires emerge more or less at the same time.

> **Landmark number** Earliest writing—3200 BCE

5000–4000 years ago (3000 BCE to 2000 BCE)

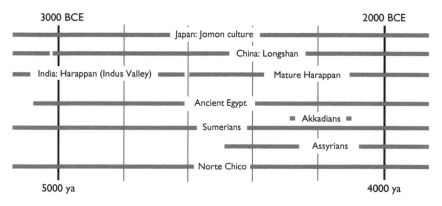

It's estimated that there were 45 million people in the world in 3000 BC. Having made the case for the importance of writing, we have no written evidence for the very first of the great cultures that we know about.

- In Japan, the Jōmon hunter-gatherer culture, which left many pottery artefacts, has been dated to 14,000 BCE.
- In the Yellow River area of China, there is evidence of a coherent, extensive and long-lasting culture dubbed 'Longshan' from 3000 BCE to around 1900 BCE. The beginning of the reign of the legendary 'Yellow Emperor' is dated to around 2700 BCE.
- In India, the Indus Valley civilisation flourished between 3300 BCE and 1300 BCE.[55]
- In Ancient Egypt, one of the sites of the earliest writing, there is a continuous culture from before 3000 BCE right through three millennia to the start of the Common Era, when the Romans defeated Cleopatra and Mark Antony at Actium. The Great Pyramid of Khufu was built between 2580 and 2560 BCE.
- In the Middle East, the Akkadians[56] and the Sumerians flowered and fell. The Assyrian Kingdom arose in around 2500 BCE.
- In South America, the Norte Chico civilisation flourished in what is now Peru. This is the oldest known civilisation in the Americas and lasted until around 1800 BCE.

In Eurasia, this age of early empires falls squarely into the Bronze age. By 2000 BCE, it's estimated there were 72 million people alive.

4000–3000 years ago (2000 BCE to 1000 BCE)

- In China, the first of the traditional dynasties, the Xia, arose roughly in the area where the Longshan were, around the Yellow River. They were succeeded by the Shang Dynasty around 1600 BCE, and the earliest Chinese writing dates to the middle Shang Period. The Shang were in their turn supplanted by the Zhou Dynasty by 1000 BCE.

---

[55] The Indus Valley civilisation **did** leave a written legacy. Unfortunately, we've not yet been able to decipher their writings.

[56] The Akkadians have a claim to being the first Empire in history and occupied the fertile areas reaching from present-day Syria, through Iraq, down to Turkey. They had a somewhat symbiotic relationship with Sumeria, where the very earliest evidence of writing is found.

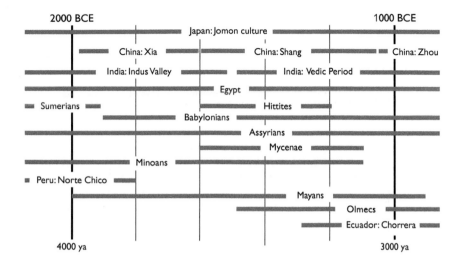

- In India, the Vedic Period (1500–500 BCE) succeeded the Indus Valley civilisation. It takes its name from the Vedas, Hindu scriptures composed at that time.
- The Middle East was dominated by the Assyrians and Babylonians in what is now Iraq, and the Hittites in modern-day Turkey.
- The Mediterranean region saw the rise of some notable civilisations, Minoans in Crete, who left behind a written legacy in the Linear B script, and the Mycenaeans, possibly the Greeks of Homeric legend. Some date the fall of Troy to around 1200 BCE.
- In Middle America, from around 2000 BCE, we find the Mayans, a culture that was to last for 3½ thousand years. The Olmec culture arose in around 1500 BCE, and would last for around 1000 years.
- In South America, the cultures known as Quitu (from around 2000 BCE) and Chorrera (from around 1300 BCE) arose in the region of modern-day Ecuador and lasted around a thousand years. The Chorrera left behind a wealth of ceramic artefacts.

The Bronze Age ended abruptly, between 1200 BCE and 1150 BCE, with the simultaneous collapse of cultures in Mycenae, Anatolia, Syria, and other places. Palace cultures reverted to village cultures.

By 1000 BCE, the world's population had grown to an estimated 115 million.

Landmark number Collapse of Bronze Age cultures—1200 BCE

3000–2500 years ago (1000 BCE to 500 BCE)

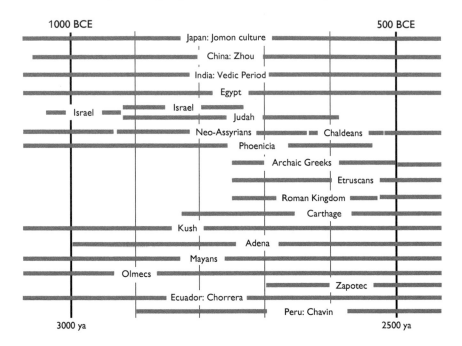

- China saw the rise of the Zhou Dynasty from 1046 BCE, which proved to be the longest-lasting of the Chinese dynasties—almost 800 years. The thinker Laozi's life falls into this period, and Confucius was alive at the end of it. The Zhou's authority over other states declined from 770 BCE, in what is known as the Spring and Autumn Period.
- India: The late Vedic Period saw a transition from nomadism to settled agricultural communities.
- Middle East: The Iron Age empires of the Neo-Assyrians, Neo-Babylonians (Chaldeans), and Persia (Achaemenids) rose to prominence. This was also the time of the Kingdoms of Israel and Judah.
- Mediterranean: The Phoenicians developed trade throughout the Mediterranean. The classic Greek civilisation was forming.
- Italy: The Etruscans flourished in what is now Tuscany, while around 750 BCE the inhabitants of a small swampy town called Roma began to throw their weight around.
- North Africa: Carthage was founded by Phoenicians in 814 BCE.

- East Africa: The city of Meröe was the centre of the Kingdom of Kush, in the region known as Nubia, south of Egypt. Initially conquered by Egypt, the Egyptians had withdrawn by 1100 BCE, and the Kingdom of Kush would last for over 1200 years.
- North America: A culture known as the Adena flourished across a wide area centred on the Ohio River basin.
- Middle America: The Olmecs and Maya continued. A civilisation known as the Zapotec emerged around 700 BCE.
- South America: The Chavin culture dates from around 900 BCE and arose in the Andean highlands, but was influential also along the Pacific coastline in what is now Peru.

By 500 BCE, the population of the world had reached 150 million people.

2500–2000 years ago (500 BCE to 1 BCE)

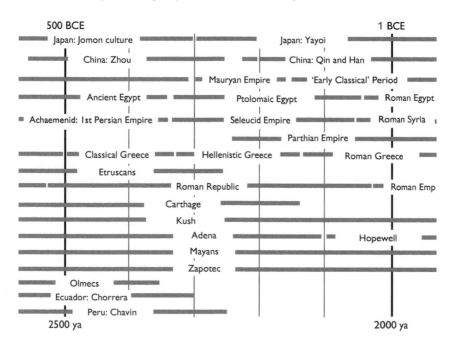

- Japan: The Yayoi Period lasted for about 600 years from about 300 BCE, and saw the start of intensive rice cultivation.

- China: The Zhou Dynasty fell in 256 BCE after approximately 250 years known as the Warring States Period. The Zhou were replaced by the first Imperial dynasty, the Qin,[57] who were rapidly succeeded by the Han. The Han Dynasty especially is regarded as a golden age of imperial China.
- India: Siddhārtha Gautama (Buddha) was born around the start of this period in the Magadha Kingdom—estimated dates of his birth range from 563 BCE to 480 BCE. The Mauryan Empire arose in around 320 BCE, lasting just 140 years, and among its rulers was Ashoka, who embraced Buddhism. The Satavahana Empire dominated central India in the second half of this period, while the Pandyan Empire flourished in the south. The Mauryan Empire ended in 184 BCE.
- Middle East: The Persians dominated the start of this period, but were displaced by Alexander the Great and the successor Seleucid Kingdoms. The Parthians conquered Mesopotamia and surrounding areas in around 250 BCE, creating an empire that would last almost 500 years to 220 CE.
- Around 5 BCE, Jesus of Nazareth was born in Roman Judea.
- Mediterranean: In Greece, the Greeks defeated the Persians before starting a civil war, the Peloponnesian War. Greece later fell under the sway of Alexander the Great, but by the end of this period was part of the Roman Empire.
- Carthage was established as a state independent of the Phoenicians by 650 BCE and, after a long conflict with Rome, was defeated in 146 BCE.
- Italy: Over this period, Etruscan cities were gradually absorbed by the Romans, who were expanding their territory inexorably. Eventually, in 27 BCE, the civil war and political convulsions surrounding Julius Caesar's rebellion and assassination gave birth to the Roman Empire.
- East Africa: The Meroitic Kingdom (Kush) continued through this period.
- North America: The Adena culture declined, and in more or less the same territory, the Hopewell tradition arose, consisting of a group of associated populations connected by a network of trade routes.
- Central America: By now the Mayans had become the first culture in the Americas to use writing in the form of a hieroglyphic script.
- South America: This period saw the decline of the Chorrera and Chavin cultures in Ecuador and Peru.

[57] The first emperor, Qin Shi Huang, was responsible for the famed terracotta army buried with him. His name, and that of his dynasty, is the origin of the name China, and was formerly written as Ch'in in Roman characters.

> **Landmark number** Conquests of Alexander the Great—330 BCE

When it comes to empires, all roads do seem to lead to Rome. The Roman Republic lasted approximately 500 years—and the Roman Empire (the Western part after the split between East and West) also lasted approximately 500 years. Very neatly, the transition happened in 27 BCE,[58] more or less at the zero point of our conventional year numbering system.

By 1 CE, the population of the world had reached 188 million people.

2000–1500 years ago (1 CE to 500 CE)

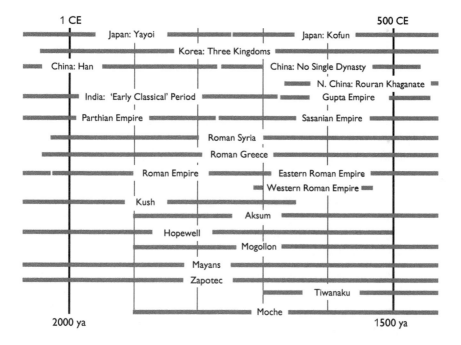

- Japan: The Kofun Period, which followed the Yayoi Period, is the earliest part of the Yamato Period, starting from around 250 BCE. It marks the start of Japanese recorded history, the earliest known Japanese writing dating to the fifth century CE.

---

[58] 'Before the common era'. There is actually no zero point for our year numbering system. 1 BCE is followed immediately by 1 CE, an annoying anomaly.

- Korea: The Three Kingdoms of Korea were the Silla, which lasted for almost a thousand years, starting in 57 BCE, and the Goguryeo and Baekche Kingdoms, which were taken over by the Silla in around 660 CE.
- China: The Han Dynasty continued, lasting until 220 CE, when it was succeeded by a series of shorter-lived dynasties, the Three Kingdoms, the Jin, and the period known as the Northern and Southern Dynasties. The Rouran Khaganate was a proto-Mongol state in Northern China from around 330 CE, and lasted 220 years.
- India: The Satavahana Empire continued until 220 CE, while the Pandyan Empire continued to rule in the South. The Kushan Empire arose in areas stretching through Afghanistan into Northern India, falling in 375 CE to the Sasanians from the West and the Guptas from the East. The Guptas were dominant in India from 320 CE to 550 CE, a period regarded as a golden age for India.
- Middle East: With the Parthians in decline, the Sasanian Empire, the last flourishing of the Persians prior to the rise of Islam, came to power in around 200 CE, and ruled until 651 CE in an area covering the Middle East and stretching through to Afghanistan and Pakistan.
- Mediterranean, Italy, and much of the rest of Europe: The Roman Empire. What more is there to say? From the accession of the Emperor Augustus in 27 BC to the deposition of Augustulus in 476 that marked the end of the Western Roman Empire, the Romans dominated the Mediterranean and surrounding lands for 500 years, and continued in the East for almost another thousand years.
- Africa: The Meroitic Kingdom of Kush ended by 350 CE. The trading nation of Aksum, in Northern Ethiopia, lasted from 100 CE for more than 800 years.
- North America: The Mogollon culture flourished in the southwestern part of North America from around 200 CE until conquered by the Spanish in around 1500 CE.
- Central America: At this time, Teotihuacan, in modern-day Mexico, was the largest city in the Americas. The Mayans continued, as did the Zapotec culture.
- South America: The Moche civilisation arose in Northern Peru. From 300 CE, the Tiwanaku Empire grew along the coast in what is now western Bolivia.

> **Landmark number** Fall of the Roman Empire in the West—476 CE

It's estimated that by 500 CE, the world's population was 210 million people.

1500–1000 years ago (500 CE to 1000 CE)

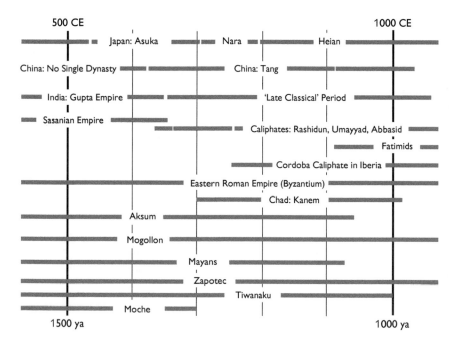

- Japan: The Yamato Period continued through a series of sub-periods: the Asuka, notable for the introduction of Buddhism, the Nara, and the Heian Periods.
- China: The period known as the Northern and Southern Dynasties ended with a unification under the Sui Dynasty, which lasted just 37 years and was succeeded by the Tang Dynasty, seen as a golden age and lasting almost 300 years. After a period of upheaval known as the Five Dynasties and Ten Kingdoms, the Song Dynasty was the dominant power by 1000 CE.
- India: The Empire of Harsha lasted from 606 CE for about 40 years, and was notable for its cosmopolitan nature. It was succeeded by the

Chalukya Dynasty for 110 years, and then by the Rashtrakutas, a period known as 'late classical'.

- The year 622 CE—the date that Muhammad travelled from Mecca to Medina—marks the beginning of the Islamic calendar and of a series of changes that transformed the region. The years that followed saw the rise of Islamic caliphates over large areas of the Middle East, Arabia, and North Africa. First was the Rashidun Caliphate, which lasted around 30 years, followed by the Umayyad Caliphate for around 90 years to 750 CE. From that date, the Abbasids were dominant, with the Fatimid Shia Caliphate powerful in North Africa and Arabia.

- Iberia: The Islamic Umayyads (also known as Moors) moved into Iberia in the early eighth century and governed Al-Andalus from 800, co-existing with Christian states, who were mainly in the North.

- The old Eastern Roman Empire, also known as the Byzantine Empire, was powerful in Europe until 700 CE, after that contracting to the Middle East and dwindling after 1100.

- In Africa, the Kanem Empire was centred on what is now Chad, and lasted more than 650 years from around 700 CE.

By now the world's population was approaching 300 million, more than six times what it had been 4000 years earlier. And in the thousand years to follow, that number would grow more than twenty-fold.

Multidimensional Measures

Areas and Volumes

If you know how many acres you have sown of each kind of corn, inquire how much the acre the soil of that land takes for sowing, and count the number of quarters of seed, and you shall know the return of seed, and what ought to be over.

Robert Grosseteste

Which of these has the least volume?
☐ Volume of water in the Fort Peck Dam (USA)
☐ Volume of water in Lake Geneva
☐ Volume of water in the Guri Dam (Venezuela)
☐ Volume of water in the Ataturk Dam (Turkey)

Squares for areas and cubes for volumes

Them!

In the 1956 horror movie, *Them!* (the exclamation mark is part of the title), the plot revolves around 'atomic testing in 1945 [that] developed...dangerous mutant ants'. Naturally, giant ants escape, wreak devastation and threaten life and limb. But why don't we actually see ants of that size on Earth? Is there some reason that the largest ants in the world, bullet ants from Panama, have evolved to a maximum length of 'only' 4 centimetres? The answer lies in the numbers and how measurements of lines, areas, and volumes relate to one another.

We've looked at length, the fundamental measurement of space, but remember that there are three 'length' dimensions that measure the space we inhabit.

And that means dealing with squares for areas and cubes for volumes, and those can produce challengingly big numbers very rapidly.

Covering the ground

You've probably seen it in dozens of television crime dramas. A row of searchers, police and volunteers, walk methodically across a moor, looking left and right, scanning for signs of a missing person. Let's put some numbers into that picture.

Suppose, to start, that the area to cover is just 100 metres by 100 metres.[59] Roughly the size of Trafalgar Square in London, or a football pitch. Let's imagine a row of 25 searchers, spaced 4 metres apart, and they're lined up along one edge of that square—each searcher must look 2 metres to their left, 2 metres to their right. And let's imagine they can progress at rate of one metre per second. Then it will take 100 seconds to sweep that patch of land—1 minute and 40 seconds. Hardly worth bringing 25 people out for that, but this is just a thought experiment!

Now suppose the area to cover is 1 km by 1 km—ten times wider, ten times longer. That's still a relatively modest area to cover, but for the same team of 25 searchers it will need 10 passes, and each pass will take 10 times as long. So in all it will take 100 times as long. That's 10,000 seconds—around 2¾ hours, working without a break. Increase the search area to 10 km by 10 km and the time taken to do the job has grown to a million seconds—that's more than 11 days, working around the clock.

When we move from thinking about distance to thinking about areas, we go from one dimension to two, and that makes all the difference. Pity the spotter plane searching the open ocean for signs of a downed airliner.

Our brains need help when thinking about areas because so much of our early learning about numbers and measurements is linear. There are no tools in the set of school mathematical instruments to measure area: areas are almost always the result of calculations, and are seldom measured directly. Think of the specifications of a computer monitor. You might think that the relevant metrics for such a two-dimensional device would include the area of the display, or indeed the total number of pixels. But no, the technical specs invariably refer to the diagonal measure (for overall size) and the horizontal and vertical pixel counts separately, for resolution.

But two-dimensional measures are called for whenever we need to cover a surface, whether it's to search the moors, pave a driveway, sow a field, carpet

---

[59] That's one hectare, the metric unit of areal measurement, equal to about 2½ acres.

a bedroom, or just paint the kitchen. The fact is, that when it comes to areas, the numbers involved get very big very fast.

The equator is 'just' 40,000 km long: the surface area of the Earth is 510 million square kilometres (km²), a number four orders of magnitude bigger than the linear measure of a great circle. There's no deep significance there, it's plain geometry and arithmetic, but the fact is that when dealing with areas and volumes, the numbers are apt to be big.

> **Landmark number** The surface area of the Earth (land and sea) is just over 500 million square kilometres, or, if you prefer, half a billion.

Speaking (of) volumes

And if the shift from linear to areal measures is hard to fully grasp, it gets even trickier when it comes to volumes.

Let's flex our **visualisation** skills a little. It's estimated that 2.3 million blocks of stone make up the Pyramid of Khufu[60] in Egypt. Is that a plausible number? Let's see how we can make sense of that big number, and perhaps visualise it in a way that is both comprehensible and memorable.

If those blocks were all cubes of the same size (they're not, but this is just an exercise), how many blocks would there be on each side of an idealised pyramid (of roughly the Khufu's proportions) containing 2.3 million blocks?

Well, a little bit of pen-and-paper (and calculator) playing around with the numbers, and using the formula for the volume of a pyramid,[61] reveals that to match that total cubic measure, a pyramid of the right height-to-base-length proportion would need to measure 225 blocks × 225 blocks on the base and be 136 blocks high (and we can calculate that that would give a total of a little under 2.3 million blocks). The actual dimensions of the great pyramid of Khufu are approximately 230 m × 230 m × 139 m. These numbers are eerily close to our counts of blocks. This means that, if the blocks were all equally sized cubes, the average size of those cubes would be approximately 1.022 m on each side, and that seems perfectly plausible.[62]

---

[60] Also known as the Great Pyramid of Giza, or the Pyramid of Cheops

[61] To calculate the volume of any shape that 'tapers off' to a point, like a pyramid or a cone, multiply the base area by ⅓ of the overall height. So in this case we have 225 × 225 × 136/3 = 2,295,000.

[62] Of course, the blocks are not all the same size, nor are they all perfect cubes. And we haven't allowed for the chambers inside the pyramid, and there are many other objections that could be raised. But this is a reasonability check, not a quantity surveyor's report. We're after the answer to the question '2.3 million blocks: is that a big number?'

More than plausible, it's almost uncanny that the average size of the blocks of the pyramid are so very close to matching the historic standard measure of a yard, and the modern standard measure of a metre!

So although 2.3 million blocks is indeed a big number, one we struggle to mentally grasp, when we express it as 225 × 225 × 136/3 we can bring it down into our mental comfort zone, and we can easily see that it's a credible estimate of the number of blocks used to build the pyramid.[63]

Going back to Earth-measuring, we saw before that the Earth's circumference is in the 40,000 kilometres range, and its surface area is around 500 million square kilometres. How big is its volume? 1.08 trillion cubic kilometres. A big number indeed.

No need to fear the giant ants

The relationship between length, area, and volume is sometimes called the 'square–cube law': as the linear dimension of an object changes, so the surface

---

[63] Continuing the reasonability checking: if the blocks are of limestone, which has a density of approximately 2500 kg/m³, then each block will weigh around 2670 kg, meaning that the pyramid as a whole will have a mass of around 6 billion kilograms. Wikipedia suggests the pyramid has a mass of 5.9 billion, which seems quite plausible.

area increases by the square of the multiplier, and the volume increases by the cube of the multiplier.

The square–cube law explains why the giant ants of the movie *Them!* are so improbable. From the movie poster, these ants look to be easily four metres in length, which means they must be around 100 times longer than even the Panamanian bullet ants.

But, in accordance with the square–cube law, a hundred-fold increase in length would mean a ten-thousand-fold increase in measures of area and a million-fold increase in measures of volume (and therefore mass). Since the strength of the ants' legs would be related to the **area** of the cross-section of their limbs, while the mass of their bodies would relate to their **volume**, it follows that the ants' bodies would now be a hundred times too heavy for the strength of their limbs, and they would simply collapse under their own weight. The same would apply to the mass of their internal organs, now a hundred times too heavy to be contained by their chitinous 'skins'. Visualisation of the messy details is left as an exercise for the reader.

This also explains why the bigger animals get, the more thick-set they have to be. A gazelle scaled up to the height of an elephant would snap its delicate legs. An ostrich cannot fly: there is a size limit to flying creatures, and this is all governed by the laws of squares (the area of the wing) and cubes (the mass of the creature).

String theory

One final example: take a look at this sequence of size versus mass for four different orchestral string instruments:

Instrument	Length	% of violin	Mass	% of violin
Violin	0.6 m	100%	0.4 kg	100%
Viola	0.69 m	115%	0.54 kg	135%
Cello	1.22 m	203%	3.5 kg	875%
Double bass	1.9 m	317%	10 kg	2500%

If these were all solid bodies, we'd expect the mass to increase proportionately with the volume of the instrument,[64] or the third power of the length. And, if

---

[64] No, I don't mean volume as in loudness.

they were entirely hollow bodies made from wood of consistent thickness, we'd expect the mass to increase with the square of the length. But in fact the progression of masses falls between the 'volume' and the 'area' scales. Obviously, a double bass is not just a scaled-up violin: the proportions of the parts will change as the size increases, but this again shows how, when things get bigger, the mass and volume increase disproportionately.

Land areas

Happy the man whose wish and care a few paternal acres bound, content to breathe his native air in his own ground.   **Alexander Pope**

Tell her to find me an acre of land

Land has always been a scarce resource, and that makes it valuable. It's no surprise then that the measuring of land goes back to the earliest civilisations.

The story of the founding of Carthage tells of Queen Dido bargaining for land with the Berber King Iarbas in North Africa, gaining agreement for only as much land as could be encompassed by an oxhide. Cunningly, Dido cut the oxhide into very fine strips (exploiting the bit of the square–cube law that deals with area), and then in a further twist she proceeded to enclose, not a circle of land, but a semicircle bounded on one side by the sea. Mathematicians know this question of how to enclose the greatest possible area between a curve of given length and an unlimited straight line as 'Dido's Problem'.

Much of ancient Egyptian mathematics involved the measurement of land and the laying out of plots.[65] Since the annual flooding of the Nile regularly obliterated property boundaries and changed land shapes, the skills of the surveyor were in perennial demand. The Egyptians used a basic unit of land area called an *aroura* or a *setjat*, which was defined as a square with each edge measuring 100 royal cubits (about 52 m long), which gave an area roughly 2700 m^2 in metric terms.[66]

The Romans measured land using a unit called the *iugerum*, which Pliny the Elder defined as 'the amount of land that a yoke[67] of oxen can plough in a day', and

---

[65] Including the use of a rope marked off in sections of lengths 3, 4, and 5, which when stretched made a right-angled triangle.

[66] A soccer pitch is a little more than 7000 m^2. So this is around ⅜ of a soccer pitch.

[67] Etymology of yoke? From the Latin *iungere* 'to join', also the root of junction and jugular. Of course, this is also the root of *iugerum*.

it measured 240 (Roman) feet long by 120 feet wide,[68] equivalent to 2523 m². This is remarkably close to the size of the Egyptian measure, which may suggest that the Egyptian measure was also derived from a practical amount of land to be ploughed in a day.

The Romans called two *iugera* a *heredium*, and this was the amount of land traditionally said to have been granted by Romulus to each citizen, and the largest amount that could be passed on as an inheritance.

The mediaeval European definition of an acre[69] also was the amount of land that could be ploughed by a span of oxen in one day. This folk definition is not a precise measure—but it was formalised as a rectangle of land 1 furlong (220 yards) in length and 1 chain (22 yards) wide, and in modern terms equal to 4,047 m². The folk definition of the furlong was the length of furrow that a span of oxen could plough without needing to take a rest. So it's not hard to visualise the acre of land, with the ploughman resting the oxen at the end of each furlong before starting the return journey. You may note that, as the acre is fully 60% larger than the iugerum, it seems the mediaeval oxen had to work harder than the Roman ones!

In the English mediaeval system, fifteen acres made an 'oxgang'; eight oxgangs made a 'carucate', and a carucate was the amount of land a team of eight oxen could plough in a season.

My father, among his many roles as a country lawyer in South Africa, dealt with the sale of farms. And until metrication in the 1970s, I can remember him talking about the areas of farms in terms of *morgen*, a unit inherited from the Dutch colonists, and also used (in various forms) in Germany and a handful of other countries. The word *morgen* means 'morning', and was notionally the amount of land that could be ploughed in the morning hours of the day. The South African morgen-to-acre conversion was 1 morgen to 2.12 acres. If you were to take the 'amount ploughed' definition seriously that would mean that South African oxen could plough twice as much in a morning as the British ones could in a whole day!

Making land metric

The International System of units (SI) introduced after the French Revolution included a unit of land area called the 'are': 10 m × 10 m. In practice, the are is not itself much used, but the hectare, 100 ares (100 m × 100 m), became the

---

[68] A roman foot was about a third of an inch smaller than the modern measure.

[69] The word 'acre' derives from Proto-Indo-European *agro-* meaning a field.

default metric measure for land area. The hectare is about 2.47 acres and about four of the Roman *iugera*.

Many countries with traditional measures for land area simply declared these older units to be equal to a hectare, thus retaining the traditional names while gaining the benefits of international standards. Thus, the *jerib* in Iran, the *djerib* in Turkey, the *gong qing* in Hong Kong/Mainland China, the *manzana* in Argentina, and the *bunder* in the Netherlands are all by modern definition equal to the hectare in size.

Here are some guides to help you to visualise a hectare: a square circumscribing the base bastion of the Statue of Liberty measures approximately one hectare in size, as does Trafalgar Square in London—and a rugby football field is also close to a hectare in area.

Farms and other properties are conveniently measured in hectares, but for larger land areas, the numbers quickly start becoming unwieldy, and so the natural next bigger unit for measuring land area is the square kilometre—100 hectares.

> **Landmark number** A rugby football field is approximately 1 hectare in area.

City sizes

When we talk about the size of cities, we normally refer to their population, but this chapter is about land areas, and for now we're talking about the physical amount of land covered. You'd think that this should be straightforward, but it is not so. For a start, there are several ways to define the area of a city:

- Its urban area, the central core
- Its administrative area, the part controlled by the city government (if such a body exists)
- Its metropolitan area, including all affiliated commuter suburbs
- Or simply the largest contiguous built-up area, not bothering that several cities might have merged into one megalopolis

Even these definitions are somewhat ambiguous, but the metropolitan area will serve for our aim of establishing a feel for the numbers. Greater London's metropolitan area is 1569 km² in extent. One way to visualise this is as a circle approximately 44 km across. Quick reasonability check: the M25 motorway circling London does indeed have a diameter of between 40 and 50 kilometres.

New York's metropolitan area, known as the Tri-State Area, is around 34,500 km² in area, while the Greater Tokyo Area is 13,500 km².

The largest contiguous built-up area in the world (by population count) is the Pearl River Delta in China, centred on Guangzhou and including Hong Kong and Macau. It's not just big in population terms, it's a vast area of land too, some 39,380 km², equivalent to a circle of 224 km edge to edge. It would take 2 hours to drive across such a circle, given a smoothly flowing freeway.[70]

An area the size of Wales

It's a cliché in the UK to describe largish areas of land by reference to the size of Wales. There's a charity that calls itself 'Size of Wales' and has campaigned to protect an area of rainforest equal to, yes, you guessed, the size of Wales. So how big is Wales actually? It's a little under 21,000 km². For comparison, that means that the Pearl River Delta urbanised area is nearly twice the size of Wales.

Other territories are used as standard measures too. As a child taken to see wildlife in South Africa, I was told that the Kruger National Park was 'the size of Israel'. Well, is it? The size of the Kruger Park is actually 19,485 km², which is not far off, at about 94% of the size of Israel, which is 20,770 km², itself pretty much the size of Wales.

Landmark numbers
- Wales is 21,000 km² in area.
- Israel is pretty much the same size.

And if you had a heart as big as Texas, how big would your heart be? It would be 696,000 km², and that's 33⅓ times the size of Wales.

The smallest country in the world is the Vatican City at just 0.44 km², entirely enclosed within the city of Rome, but nonetheless an independent nation. The largest nation in the world by land area is Russia, at 17.098 million km², which works out to 3.2% of the Earth's surface, and 11.4% of the land area.

How big is a country, typically?

Here's a chart showing the sizes of 256 countries, ranked in order of land area, from the largest (Russia), to the smallest (the Vatican).

---

[70] But since this is a coastal area, maybe we should follow Dido's lead in claiming Carthage and visualise a semicircle bounded by the coast. In this case, we'd need a semicircle that is approximately 300 km along its straight edge, and extends inland for 150 km.

A few things to note about this chart:

First, the extent to which the largest country, Russia (17.1 million km²), outstrips all other countries for size.

Second, the pack of runners-up, made up of five very large countries. Russia is followed by Canada, China, and the USA in that order, all with between 9 and 10 million km² (counting land and inland water) and therefore each a little more than half the size of Russia. Then comes Brazil at 8.5 million km² and Australia at 7.7 million km².

Third, the sharp fall to the remaining countries, the tail. Largest of those in the tail, in 7th place overall, is India (3.3 million km²), still less than half the area of Australia. But the tail is very long and thins very quickly. Egypt is in 30th position and is just over 1 million km². Iceland is in 108th position with just over 100,000 km², around $^1/_{100}$ the size of Canada. And Iceland, small as it is, is nonetheless well in the top half of the table.

So this is an extremely skewed distribution: Yemen, whose area (528,000 km²) is very close to the mathematical average of country sizes (533,000 km²), is nonetheless in position 50 in the list. The median country size (half the countries are larger, half are smaller), is around a tenth of the average size— 52,800 km². Croatia is a little bigger (56,600 km²) than the median, and Bosnia– Herzegovina is a little smaller (51,200 km²). So, if there is such a thing as a typical country, it's small.

Landmark numbers
- Russia, largest country in the world—17 million km²
- China, most populous country in the world—9.5 million km²
- Median country size, around 50,000 km²

So Russia is 17 million km²: is that a big number? Can we **visualise** it? It's equivalent to a square that is just over 4000 km on each side. That's around 10% of the length of the equator, 40% of the pole-to-equator distance, and the East-to-West width of Australia.

China, Canada, and the USA, at something over 9 million km², would each be equivalent to a square of around 3000 km to a side.

The largest country in Africa is the Democratic Republic of the Congo at 2¼ million km², about an eighth of the size of Russia, while the largest country in Western Europe is France, with 640,000 km², less than $1/_{25}$ of Russia's size. (France would be equivalent to a square just 800 km on each side, though since the French like to see their country as shaped like a hexagon, we can visualise a hexagon with 500 km on each side.)

The United Kingdom measures in at 242,000 km². That would equate to a square somewhat less than 500 km on each side.

Measuring continents and large islands

The concept of 'continent' is ill-defined. Should Europe and Asia be treated as separate continents when they are clearly part of the same land mass? What about Australia? Continent or island? Some classifications even regard 'the Americas' as a single continent. We'll use the commonly used seven-continent scheme, and that gives us the following sizes of land area:

Asia	43,820,000 km²
Africa	30,370,000 km²
North America	24,490,000 km²
South America	17,840,000 km²
Antarctica	13,720,000 km²
Europe	10,180,000 km²
Australia/Oceania	9,008,500 km²

And the largest islands in the world are:

Australia (on its own)	7,692,000 km²
Greenland	2,131,000 km²
New Guinea	786,000 km²
Borneo	748,000 km²

[Texas would come here if it were an island: 696,000 km²]

Madagascar	588,000 km²
Baffin Island	508,000 km²
Sumatra	443,000 km²
Honshu	226,000 km²
Victoria Island (Canada)	217,000 km²
Great Britain	209,000 km²
Ellesmere Island (Canada)	196,000 km²

When judging land areas, world maps can be very deceptive: the commonly used Mercator projection suffers from the flaw that it seriously distorts areas, greatly magnifying areas closer to the poles.

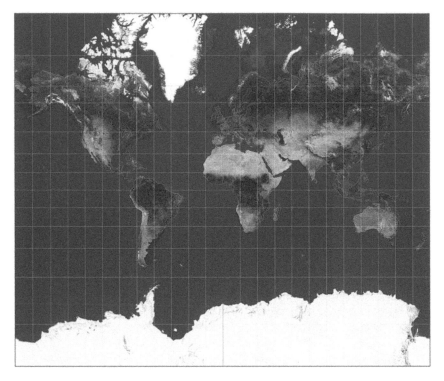

Map created by Strebe and hosted on Wikimedia Commons, reproduced here under a Creative Commons licence.

For example, using a Mercator projection as shown above, the apparent sizes of Greenland and Africa on a map are very similar. However, looking at the numbers above, we can see that Africa is more than 14 times the size of Greenland.[71]

Adding together all the land area of the Earth, we get to around 149 million km². The rest is water, and adding in 361 million km² for all the seas and lakes, we get to 510 million km² total surface area.[72]

---

[71] The Mercator projection is excellent for navigation: lines of latitude and longitude are parallel, and directions of travel are accurately shown. With flat maps of the world, you have to choose which qualities you need to be accurate and which can be sacrificed. Only a globe truly succeeds in mapping the Earth accurately.

[72] And taking the radius of the Earth as 6370 km, we can calculate the same number, 510 million km², using the formula for the surface area of a sphere.

Sipping and shipping: measuring volume

Roll out the barrel

When you have a valuable product, you keep careful track of it. Through the centuries, alcoholic drinks have been very carefully transported, stored and dispensed in measured quantities.

In Elizabethan England, there was in use a sequence of measures for liquids based on a doubling principle. The largest quantity of wine measured by the system was known as a tun, and was at the head of this marvellous chain of capacities:

A tun was 2 pipes,
    a pipe was 2 hogsheads,
        a hogshead was 2 barrels,
            a barrel was 2 casks,
                a cask was 2 bushels,
                    a bushel was 2 kennings,
                        a kenning was 2 pecks,
                            and a peck was 2 gallons.

A gallon was 2 pottles,
    a pottle was 2 quarts,
        a quart was 2 pints,
            a pint was 2 cups,
                a cup was 2 gills,
                    a gill was 2 jacks,[73]
                      a jack was 2 jiggers,
                        and a jigger was 2 mouthfuls.
And a mouthful was about 1 cubic inch!

This made a tun equal to 2^{16} or 65,536 mouthfuls.[74] I love the neatness and the sheer doggedness of this system (not to mention its foreshadowing of the

---

[73] Jack and Gill? We've met them before!

[74] Let's do a reasonability check: is this a big number? What would a cuboid of volume equal to 65,536 cubic inches look like? Given this doubling sequence, one **tun** was 2^{16} cubic inches. We can

binary/doubling schemes of modern measurement of computer data). This was not by any means an exhaustive list of the liquid measures in use. Over the years, many alternative names for these volumes were used: for example, the British Navy dispensed a daily 'tot' of rum, at one-eighth of a pint, equivalent to the jack in the scheme above, and 70 ml in modern metric measures.

A variant of this doubling-up scale was even proposed in 1967 as a new modern standard in the USA by engineer Harold Larson (an opponent of the metric system). It was, of course, not taken up.[75]

Wine bottles, too, have been named in a bewildering variety of 'standard' ways, and they too sport a doubling sequence. In particular, for champagne:

A standard bottle is 0.75 litres,

    2 standard bottles is a Magnum,

        2 Magnums is a Jeroboam,

            2 Jeroboams is a Methuselah,

                and 2 Methuselahs is a Balthazar—16 bottles.

Sadly, there the binary sequence breaks down. Bigger, named, bottle sizes do exist ('Goliath' is 36 bottles), but these are not part of any unified doubling structure, often having been created for particular promotional purposes.

Oil, by the barrel or by the tankerful

Alcohol is not the only liquid of great commercial value, nor the only liquid conventionally shipped in barrels. If there is a liquid that truly underpins our modern lifestyle, it is oil.

Oil prices are quoted in terms of dollars per barrel. So we know that there must be a standard oil barrel size. How big is that measure? In fact, there is some variation in usage, but the most common definition is the US 42-gallon barrel of just under 159 litres. According to the American Oil and Gas Historical Society:

A 42-gallon tierce weighed more than 300 pounds – about as much as a man could reasonably wrestle. Twenty would fit on a typical barge or railroad flatcar. Bigger casks were unmanageable and smaller were less profitable.

rewrite this as $2^{16} = 2^6$ in $\times$ 2^5 in $\times$ 2^5 in. So this would be equivalent to a cuboid 64 inches tall and 32 inches on each side. That's a big vessel, but not outrageously so, and will contain about (10% more than) a **ton** weight of water (or wine).

[75] Larson's smallest measure was the tablespoon, and indeed a standard culinary tablespoon is more or less equal to a cubic inch, or mouthful. And a jigger, as a unit of measure for strong spirits, is not so different from the modern 'shot' of 25–35 ml (UK common usage).

A standard barrel or drum is 0.876 m tall (that's a little less than the height of a tennis net) and 0.597 m at the widest points (the ribs used to strengthen the structure of the steel barrel).

A typical small car's petrol tank (50–70 litres) can therefore hold somewhere between a third and a half a barrel of petrol.[76] Large tanker trucks used for delivering liquids such as refined petroleum to filling stations can typically carry between 20,000 and 40,000 litres, of the order of 200 barrels, sufficient for perhaps 500 cars to be refuelled.

The largest oil tankers (ships) can carry 2 million barrels of oil. That's a capacity of the order of 10,000 times greater than a typical road tanker.[77] In the catastrophic oil spillage caused by failures at BP's Deepwater Horizon undersea well, the amount of oil discharged was estimated at 4.9 million barrels, making it around 2½ times the capacity of an oil tanker.

And to cap off this oily journey, the largest oil storage facility in the world is at Cushing, Oklahoma, in the USA, which can accommodate over 46 million barrels: that's 23 large oil tankers' worth.

At this point, the question we all should have in mind, is how much of the stuff is left underground? There's no definitive answer to that, but a reasonable estimated minimum is 1600 billion barrels economically recoverable with current technology[78]—800,000 supertankers' worth. At an estimated current usage rate of 90 million barrels (45 supertankers) per day, that's not a big number. It works out to just under 50 years' worth.

In the later chapter on energy, we'll look at just how much energy is contained in that fuel, and how alternative sources of fuel compare.

Dry goods

Shipping and carrying is not just for liquids, and while consignments of solid substances and items tend to be measured by mass/weight rather than volume,

---

[76] Of course, petrol is not the same as crude oil: in fact, a barrel of crude will yield about 45% of its volume of petrol/gasoline, and around 25–30% of its volume of diesel. The remaining 25–20% will be other oil-derived products such as jet fuel, LPG, and so on.

[77] Plausibility check: If a tanker vessel can carry 10,000 times as much as a road tanker, we'd expect its linear dimensions to be larger by a factor of the cube root of that number, around 20 times bigger. The VLCC (very large crude carrier) category is around 300 m long, and this checks out with road tankers of around 15 m.

[78] If and when the oil starts to run out, market mechanisms should drive the price up, causing alternatives to become more economically viable, and at the same time making more of the reserves economic to extract. I say 'should'. Currently the oil price is absurdly low and does not reflect the true costs, either of extraction or of the consequences of burning the stuff.

some cubic measures of volume used for dry materials are still of interest. So here's a collection of curious old-time units for measuring loads of goods:

- **Bushel**: This ambivalent unit has variously been a unit of capacity and mass. In terms of capacity, it is equivalent to 8 gallons, whether US style or imperial.
- **Hoppus**: A practical measure of the amount of usable timber in round logs. It's named for Edward Hoppus, who published a manual of practical calculations in 1736,[79] based on measuring the quarter-girth of trees.
- **Cord**: A cord is a measure of a standard load of firewood[80] that can be stacked as a cuboid 4 feet wide, 8 feet long and 4 feet high, which works out to around 3¾ m³. The name is thought to relate to the use of a standard length of cord used to judge the size of the load.
- **Stere**: A stere is the metric unit introduced to replace the imperial cord, and is a stack of firewood 1 m × 1 m × 1 m, more or less equal to a quarter of a cord.
- **Twenty-foot equivalent unit (TEU)**: Essentially a container-load of goods. Standardised containers were introduced globally in the late 1960s, and the 20-foot unit has become the smallest of the standard loads. 1 TEU is equal to 38.5 m³. An ultra-large container vessel (ULCV) is a container ship that can carry more than 14,500 TEUs and hence over 500,000 m³ of goods.[81]

It's a hard rain a-gonna fall

In the farming community in the Eastern Cape region of South Africa, where I grew up, a perennial topic of conversation was the amount of rain that the countryside was receiving. Drought was always a threat, and a conventional conversation opener in good times would be a comment that 'the dams are

---

[79] The Hoppus measurement allows for the wastage inevitable when round logs are milled into rectangular cross-sections. It's equivalent to assuming that the proportion of usable wood is $\pi/4$, or around 79%.

[80] My local supplier of firewood in Surrey, England supplies firewood by the 'load', which he defines as 1.2 m³, about a third of a cord. One load will provide me with 2 years' worth of fuel for the occasional hearthfire through the winter, and the occasional barbecue through the summer.

[81] Let's do a reasonability check. A vessel of this class might be 400 m long × 60 m wide. To carry 500,000 m³ would therefore require containers stacked to the height of 20.83 m. Since standard containers are 2.59 m tall, this means to stack them 8 containers deep. In fact, containers may be seen stacked on ships as many as 12 layers deep.

looking full'. So even though my father was a lawyer, he always owned and used a rain gauge, and for him, a smart new one in a copper casing was bound to hit the mark as a birthday gift.[82]

Even though the amount of rain falling is, by its nature, a volume of water, we use a linear measurement to quantify it. This is because when it falls on open land (and is absorbed), we notionally divide that volume of water by the area over which it falls and we end up with a linear measurement of rainfall, such as 'two inches' or 'fifty millimetres'. That's how deep the water would be if it were falling on a flat plain with no absorption and no evaporation.

Incidentally, an equivalent height of snow (50 mm of snow versus 50 mm of rain) will weigh about $^1/_{10}$ as much, and melt down to $^1/_{10}$ as much water.

In December 2015, the Cumbrian village of Glenridding, being pummelled by Storm Desmond, received 67 millimetres of rain in 24 hours. Substantial damage was suffered by the village, but is that really a big number?

A little **visualisation**: picture a house with a 10 m × 10 m footprint. The rain that fell on the roof of this house in that storm would amount to 6700 litres of water.[83] The capacity of a large rain barrel may be up to 1000 litres—so this means overfilling such a barrel almost seven times.

In fact, that is by no means a freakishly large rainfall: on this occasion, it resulted in destructive flooding because the ground it fell on was already saturated and very little of it would have been absorbed. The mountainous terrain of the catchment area meant that the runoff was all channelled into Glenridding Beck and the stream came down in tumultuous flood, rolling boulders along with it.

In effect, the volume of water that fell had been concentrated into a single line, the river bed, and the square–cube law effectively magnified the destructive energy of the storm, with devastating consequences for the small village.

---

[82] One type of rain gauge in particular appealed to the budding mathematician in me: it was shaped as a (downward-pointing) cone. So, rather than being calibrated in a linear way (as a cylinder would be), it was based on a cubic scale: twice the depth of water in the gauge meant eight times as much rain had fallen. This made for more sensitive measurement of smaller rainfalls.

[83] Recall that a litre is a cubic decimetre (100 mm, $^1/_{10}$ m). So 67 mm of rain is 0.67 dm, the 10 m × 10 m roof is 100 dm × 100 dm, and so the total volume is 6700 litres.

Relative sizes of selected countries and lakes

Area of **Portugal** (92,100 km²) is
    4 × area of **Belize** (22,970 km²)

Area of **New Zealand** (267,700 km²) is
    2000 × area of **Christmas Island** (135 km²)

Volume of water in the **Samara Dam** (Russia) (57.3 km³) is
    2.5 × volume of water in the **Fort Peck Dam** (US) (23 km³)

Volume of water in **Lake Michigan–Huron** (8,440 km³) is
    50 × volume of water in the **Aswan High Dam** (Egypt) (169 km³)

Area of **Cuba** (110,900 km²) is
    100 × area of **Hong Kong** (1,104 km²)

Area of **Sudan** (1.861 million km²) is
    2 × area of **Nigeria** (924,000 km²)

Area of **Zimbabwe** (391,000 km²) is
    5 × area of **Serbia** (77,500 km²)

Area of **Namibia** (824,000 km²) is
    40 × area of **Israel** (20,770 km²)

Volume of water in the **Three Gorges Dam** (39.3 km³) is
    10 × volume of water in **Lake Zurich** (3.9 km³)

The Fourth Technique: Rates and Ratios

Knock 'Em Down to Size

Sometimes we can tame a big number by comparing it with a known landmark number in a very specific way: by dividing it by that number, and thereby creating a ratio. This can reduce the monster to a number that is well within our comfort zone, and then it becomes much easier to grasp the importance of the beast, and judge how big or small it really is.

For example, to help understand population changes in a country, you would tend to look at birth and death rates, and not the raw count of births and deaths. To get a birth rate, count births on one hand, and the total population on the other hand, and then divide births by the population. It's similar with the measurement of speed: using your units of choice, you measure distance travelled and divide by time taken. And to compare batters in baseball, you might look to a batting average: number of hits divided by number of balls faced or at bats.

These ratios make big numbers much easier to deal with, because the first step in the comparison has already been done. You've removed a confounding factor, the size of the base. A number that is expressed as a rate has in effect been standardised as a way to make comparisons easier. For example, a big country like France will have more births in a year (824,000 in 2015) than a small country like French Polynesia (4300 in the same year). That's only natural, as the population of France is so much greater than that of French Polynesia. But the raw comparison of the count of births is not a helpful comparison. It might be more useful to standardise the measurement by expressing the number of births as a birth rate (the number of births divided by the population). When you do this, you see that in France, in 2015, the birth rate was 12.4 per thousand, whereas in French Polynesia it was 15.2 per thousand.

This trick of calculating a rate or a ratio can be used in all sorts of ways: models of car can be judged on their fuel consumption, measured as litres per hundred kilometres (or indeed miles per gallon); population density can be compared between countries if we measure it as people per square kilometre; medical provision can be measured as patients per doctor; website effectiveness can be measured as sales divided by number of website visits; and so on. All of these allow us to compare things that would otherwise be impossible to compare because they relate to base metrics of different sizes.

Per capita

Think about what we did with the birth rate there. We used one of the most useful tricks in the numerate citizen's bag: standardising *per capita* (literally, per head). By dividing a number by a count of some sort of base population, you turn a raw number of occurrences, as a measure of how common the thing is, into a measure of prevalence. And when you're dividing by a number of people, as per capita[84] implies, this literally brings the big number down to a human scale. For example, the GDP of Canada in 2015 was 1.787 trillion USD. Is that a big number? Well, given that there are 35.3 million Canadians, that means approximately 50,000 USD[85] per Canadian (per capita).

Contrast Canada with Mexico, which has a GDP of 1.283 trillion USD and a population of 119 million Mexicans. This makes the comparable figure approximately 11,000 USD per Mexican. The Mexican economy is somewhat smaller, but more important, the population is three times the size, and so the share of GDP for each Mexican is just a fifth of the equivalent share for each Canadian.

As we've seen in the previous chapter, the geographic and demographic sizes of countries vary hugely. But turning things into per capita measures lets us make sense of numbers that would otherwise be difficult to reconcile. This technique is indispensable when looking at national statistics, but it will also apply in other situations. For example, you might look at turnover per employee at large and small companies: the 'standardisation' of company size might throw up insights that would be masked by raw totals.

---

[84] Per capita = for each head. It seems that heads are easier to count than whole people. Likewise, 'poll tax', 'head of cattle'.

[85] I'm using USD here rather than Canadian dollars for comparability purposes.

Share of the total

In 2015, the UK government spent around 1.1 trillion dollars.[86] That works out to around 17,700 USD per capita. Spending on the defence budget in that year was around 55.5 billion USD. Is that a big number? It works out to just over $1/_{20}$ of the national budget. Percentages are just another way of expressing a share of the whole—so we can also call that 5%. Set against figures of around 35% for pensions and other benefits, and 17% for healthcare, it doesn't seem quite so large.

Compared with a defence expenditure figure for the United States of over 15%, it seems relatively modest. These comparisons would be difficult to make without being able to express the amount relative to some other quantity, in this case, total government expenditure. Not only that, but by expressing it as a proportion of a total, you can reduce the distortion caused by inflation and growth. You're seeking the best context for understanding the big number, and this is a good way of doing it.

Governments often announce spending plans with great hoopla, announcing amounts in the millions or billions, intended to impress. Often, on examination, the amounts involved look much less impressive when set in context of the whole budget. If you want to make the UK's foreign aid budget seem large, you call it 18 billion USD: if you want to make it seem small, you call it 0.7% of GDP. If you want to understand if it's a big number, look at it from all the angles you can.

Growth rates

If the question 'Is that a big number?' is often followed by 'compared with what?', an appropriate answer will frequently be 'compared with the last time it was measured'. In the search for context, sometimes the easiest comparison to hand is the previous measurement. So, a country's Gross Domestic Product (GDP) or predicted GDP is inevitably compared with the previous value to give a growth rate (which might sometimes be negative[87]). The growth rate is calculated as the change from the previous value, divided by the previous value (and

---

[86] Disclaimer: post-Brexit, the exchange rate has fallen dramatically. The figures I quote here use a historical average exchange rate of 1.6 USD to GBP.

[87] Two consecutive quarters of negative GDP growth count as a 'recession'. If a recession lasts two or more years, it becomes a 'depression'.

usually turned into a percentage). Growth rates have the additional benefit of a standardising effect. GDP growth rates for the UK can be compared with GDP growth rates for the USA, and the disparity in the size of the economies won't invalidate the comparison.

Growth rates can be oversensitive measures. If a measurement has a margin of error (and most large-scale measurements turn out to be estimates rather than exact measurements), this can result in wildly unstable growth rates. This is an effect often ignored by the media in search of good stories—a 5% fall in the crime rate might be explained as easily by random variation, either in crime itself, or reporting of crime, or in measurement of that reporting, as by a genuine fall.

Growth rates must also take into account periodic, typically seasonal, variations. Businesses would report their quarterly turnover comparing with the same quarter in the previous year, and not simply with the previous quarter.

For the same reason, inflation is measured, not in terms of the increase in the relevant price index over the month, but in terms of the increase over the previous year. A dramatic jump in prices one month will form part of the inflation rate for a further 11 months. In fact, each month's announcement of the inflation rate is only $^1/_{12}$ new news, and $^{11}/_{12}$ based on what's gone before.

Massive Numbers

Heavy-Duty Numbers for Weighing Up

You shall do no wrong in judgement, in measurement of weight, or capacity. You shall have just balances, just weights, a just ephah, and a just hin.   **Leviticus 19:35–36**

Which of these has the greatest mass?
☐ An Airbus A380 airliner (maximum takeoff weight)
☐ The Statue of Liberty
☐ An M1 Abrams tank
☐ The International Space Station

The weight(s) of history

If measurement of distance is the most basic of quantity measurements, then a candidate for second place must surely be mass.[88] And, like the historic measures of distance, the earliest measures of mass are based on everyday life. When it comes to weights and weighing, the oldest and deepest associations are to do with trading. So it's no surprise that the earliest basic unit of mass, in many parts of the world, is the 'grain', which in England was notionally the weight of a single kernel of barley. A different seed was used as a standard in the weighing of gold and silver: the 'carat' represented the weight of a carob seed.

---

[88] This footnote is the only place where I'm going to be pedantic and distinguish between mass (an inherent quality of something) and weight (the force of gravity acting on that thing). In general, I will use the word 'mass' unless there's a cheap word-play on offer, such as in the title of this section, or when the weight of custom and practice leaves no alternative. Frustratingly, there is no verb derived from 'mass', so we'll still have to **weigh** things to find their **mass**.

Indeed the barleycorn grain was the fundamental unit for three different systems of measurement: the Troy system, the Avoirdupois system, and the Apothecaries' system.

In the Troy system, 24 grains made a pennyweight, 20 pennyweights made an ounce (from the Latin *uncia*, one-twelfth[89]), and 12 ounces made a troy pound, for a total of 5760 grains.[90]

The Avoirdupois system (loosely translated as 'goods of weight') went through many versions, but the pound as originally defined in this system was 6992 grains (16 ounces of 437 grains[91] each). This system also introduced the unit called the stone, equal to 14 pounds. (Presumably stones were a readily available source of reference masses.) 26 stone was deemed a 'woolsack' or just 'sack', while eight stone became a 'hundredweight' (actually 112 lb, but who's counting?), and 20 hundredweight became a ton.

Where the Avoirdupois system went large, the Apothecaries' system used for medical purposes went small. Twenty grains went to make a scruple, three scruples to a drachm (pronounced 'dram'), and eight drachms to the ounce. Once again, 12 ounces to the pound.

Just as all three systems start with the grain, all of them via their various routes come to include the pound. The origin of that word is the Latin term *libra pondo*, meaning 'a pound by weight'.

From this, we get not only the English 'pound', but also the words for pound (whether referring to money or mass) in various European languages (French *livre*; Italian *lira* for money, *libbra* for weight) all of which derive from *libra*. Even in English-speaking societies, the word *libra* lives on in the abbreviations for 'pound': 'lb' for a pound of mass and '£' for a pound of money. (*Libra* also meant 'a balance scale'[92] as in the constellation in the Zodiac, and the astrological star sign in the horoscope column.)

The bushel is a curious measure, in that it has sometimes been a measure of volume or capacity and sometimes a measure of mass, sitting somewhere between the pound and the ton. Reflecting this ambiguous character, in the

---

[89] Remember that the 'inch' also derives from this Latin word.

[90] A curiosity: 240 pennyweights made a pound (weight, *libra* or lb), just as 240 pennies (money) made a pound (money). But the intermediate units (ounce for weight, shilling for money) don't sit in the progression in the same way.

[91] I know…that's an ugly number, isn't it?

[92] Okay, I lied. Here's another point about mass versus weight. The balance scale measures mass, not weight, because it *compares* the thing being 'weighed' with a reference mass. However, a scale that uses a spring measures weight, because it matches the force exerted by gravity on the object with the force exerted by the spring in tension or compression.

US system, the actual mass that a bushel signifies varies depending on the goods being weighed. So, for example, a bushel of barley is 48 lb, whereas a bushel of malted barley is 34 lb. However, both remain units of mass, and not of volume.

Ancient measures

Money and mass have an obvious connection: to a given weight of precious metal could be ascribed a specific value. So when we read in the Bible of a 'talent' as a unit of money, this derives from an underlying measure of mass. In Ancient Greece, a *talanton* was the mass of the water that would fill an amphora, and was around 26 kilograms. A talent of gold was therefore a seriously large amount of money, and it's said that a talent of silver could pay the wages of a trireme crew (200 men) for a month. The talent was subdivided into 60 *minas*, and the *mina* into 60 *shekels*.

In the biblical story of the feast of Belshazzar, Daniel reads the writing on the wall 'Mene, mene, teqel, u-farsin', and interprets this as 'measured, measured, weighed and divided'. And the word *teqel* there comes from the same root as shekel, which means 'weighing'. The shekel was further subdivided into 180 grains. The word shekel lives on, in the form of the present-day Israeli currency, the 'new sheqel'.

The link between money, weights, and trade could scarcely be plainer.

From grain to gram

As with the measurement of length, the great reform in the measurement of mass came after the French Revolution with the introduction of the metric system. Searching for a human-scale mass that coordinated with the new system for lengths, the French authorities agreed that the standard would be the mass of a one-litre volume of water (10 cm × 10 cm × 10 cm) at 4 °C, the temperature at which water is densest. A brass reference mass was made and this was designated the 'kilogram'.[93] One-thousandth of this mass (weighing as much as 1 cm × 1 cm × 1 cm of cold water) became a gram. The kilogram,

---

[93] An interesting example where the base unit is not the root word of the nomenclature. 'Kilo' + 'gram' is 1000 grams.

unique among the SI units, is still defined by reference to a physical object, a platinum–iridium mass.[94]

A metric ton ('tonne' or simply 'ton'[95]), while not formally part of the SI, is a natural extension of the SI and is used where large measurements of mass are common (as when dealing with freight), and is equal to 1000 kg. It's easy to **visualise** this as the mass of a cube of water measuring one metre on each side. To extend the visualisation, a swimming pool of dimensions 4 m × 6 m with a depth varying evenly from 1 to 2 metres would contain 36 tons of water.

So we can imagine three cubes, increasing in side length by a factor of 10 each time: the 1 cm cube weighs a gram, the 10 cm cube weighs a kilogram and the 1 m cube weighs a ton. When the linear dimension increases by 10 times, the volumetric dimension, and with it the mass of that cube (of water or anything else), increases by a factor of 1000 times. It's the square–cube law again.[96]

He ain't heavy, he's my brother

How much load can a person carry? The maximum weight that British Airways accepts as checked-in luggage is 23 kg: that's a manageable weight for a healthy adult. UK Health and Safety guidelines suggest a maximum of 25 kg for someone holding a load at waist height. The same figure is recommended as a maximum load for a backpack.

The load a person can carry depends on how that load is distributed: the 'fireman's lift' is a specific way for one person to carry another, with the carried person held over both shoulders, distributing the weight and keeping the centre of gravity of the load close to the central axis of the person doing the carrying.

Sherpas working as guides and porters for mountaineers in the Himalayas reportedly can carry loads of up to 50 kg, in some cases more than their own body weight, using a *namlo* or 'tumpline', a strap across the front of the head, which also keeps the load close to the central axis of the body.

---

[94] It looks as if this might change soon. Plans are in place introduce a new definition of the kilogram, dependent on the fundamental Planck constant.

[95] The use of 'ton' to mean a metric ton is now widespread. When I learned it at school, a ton was imperial, and was 2240 lb.

[96] And we can run this sequence the other way, too. If one millilitre of water weighs 1 gram, then a thousandth of that, a cubic millimetre, a microlitre, will weigh a milligram, about the mass of a flea. And a cube that is one-tenth of a millimetre (let's call that a hairsbreadth) on each side, will weigh a microgram. That's still easily visible to the naked eye.

And if we think just about lifting, and not carrying, in the 2004 Olympics, Hossein Rezazadeh of Iran lifted 263 kg. That's more than five times as much as a Sherpa can carry, and about equivalent to four adults, but bear in mind it's a single lift and not an extended carry.

> **Landmark number** Maximum load that a typical person can carry in their arms = 25 kg

Everyday masses

Because the mass of objects varies with the cube of the linear dimension, it can be easy to underestimate the mass of large things and to overestimate the mass of small objects. It's the square–cube rule at work again. Consider the following parade of animals:

Very roughly, the mass of a mouse (about 20 g) is
  ¹⁄₁₀ the mass of a rat (about 200 g), which is
    ¹⁄₁₀ the mass of a rabbit (about 2 kg), which is
      ¹⁄₁₀ the mass of a medium-sized dog (about 20 kg), which is
        ¹⁄₁₀ the mass of a donkey (about 200 kg), which is
          ¹⁄₁₀ the mass of a rhinoceros (about 2 tons).

So the rhino has 100,000 times the mass of the mouse, and yet in terms of body length, the rhino is only 50 times 'bigger' than the mouse (approximately 4 m as against 80 mm).[97]

Here are the masses of some human artefacts that I'd hope you can easily visualise. Starting small, to give you a feel for how small a gram is:

- 2 paper clips weigh around 1 gram (g).
- An iPhone 6 weighs in at around 170 g, an iPhone 7 at 138 g.
- An iPad Air 2 is 437 g.
- A current MacBook laptop comes to around 900 g.
- A microwave oven—around 18 kg
- A typical domestic washing machine—70 kg
- A motorbike would tip the scales at around 200 kg
- A passenger car—800 to 1500 kg

[97] Using the cube rule, we'd expect a 100,000 times mass ratio to correspond to a length ratio of 46.4. I'd say that's pretty good for a rough comparison!

- A Cessna 172 light aircraft—a mere 998 kg
- A small motorhome—3500 kg[98]
- A medium-sized truck—around 10,000 kg—'ten ton'
- A Gulfstream G550 private jet—22,000 kg
- A Boeing 737-800 jet[99]—around 40,000 kg, empty
- The International Space Station—around 420,000 kg
- An Airbus A380plus commercial jet, maximum takeoff weight—578,000 kg
- The *RMS Titanic*—52 million kg
- A supercarrier aircraft carrier—64 million kg
- *Harmony of the Seas* (world's largest cruise ship)—227 million kg

---

Landmark numbers
- Mid-range passenger car: 1000 kg = 1 ton
- Medium-sized truck: 10,000 kg = 10 tons
- Medium-sized airliner: 100,000 kg = 100 tons
- Large cruise ship: 100 million kg = 100,000 tons

Much heavier structures can be built if you don't need them to move around. Try these:

- One builder's brick—2 kg
- Eiffel Tower—7.3 million kg
- Brooklyn Bridge—13.3 million kg
- Empire State Building—331 million kg
- Burj Khalifa[100]—500 million kg
- Pyramid of Khufu—5900 million kg

Why is the Great Pyramid so much heavier than the mega-tall skyscraper that is the Burj Khalifa? Essentially because it is near enough solid stone.

How heavy was King Kong?

Thinking about the Empire State Building brings to mind that iconic image of King Kong atop the skyscraper, swatting away biplanes as he clutches Fay Wray in his massive hand. But how massive?

---

[98] This is the heaviest motorhome that can be driven in the UK without a special driver's licence.
[99] Over 4000 aircraft of this model were delivered.
[100] Currently (2017) the Burj Khalifa is the world's tallest building.

United Archives GmbH / Alamy Stock Photo

IMDb provides some relevant information on scale used in making the original 1933 movie, *King Kong*. Apparently the size of the enormous ape varies from location to location and scene to scene. The publicity described Kong as 50 ft tall, but the sets in the jungle of his home island were consistent with an 18 ft beast. The models for close-up photography of his hand were built to a scale that would fit a 40 ft animal, and the New York scenes were consistent with a 24 ft scale. Since it was the image of Kong on the Empire State Building that sparked this thought, let's go with that figure, and treat him as 7.32 m (24 ft) tall.

If we take a Western gorilla as the model for Kong when calculating height/weight ratios, we can scale up the height and use the square–cube law to scale up the weight. A very large gorilla of this species would be around 1.8 m high and would weigh about 230 kg. So Kong was just over four times as tall, and, using the cube of that ratio to scale his weight, a factor of 67.25 will give us his mass of just under 15,500 kg. Does this seem reasonable? Three times as big as an elephant? I guess so. It's entirely feasible that Kong would be scornful of the aircraft, since the planes used in the scene were Curtiss O2C-2 Helldivers, which have a gross mass of a little over 2000 kg, around one-eighth of his weight. But those annoying planes, equipped as they are with machine guns, finally cause the mighty Kong to lose his grip and tumble the 381 metres (52 times his own height) to the street below. And, compared with that iconic building, the giant ape's mass is minuscule. It comes in at less than $^1/_{20,000}$ of the mass of the Empire State Building itself.

Other creatures, great and small

100,000 kg	Blue whale—110 tons
50,000 kg	North Atlantic right whale—54 tons
20,000 kg	Humpback whale—29 tons
10,000 kg	Minke whale—7.5 tons
5000 kg	African bush elephant—5 tons
2000 kg	White rhinoceros—2 tons
1000 kg	Giraffe—1 ton
500 kg	Polar bear—475 kg
200 kg	Bottlenose dolphin—200 kg Red deer—200 kg
100 kg	Reindeer—100 kg Warthog—100 kg
50 kg	Red kangaroo—55 kg Snow leopard—50 kg
20 kg	Thomson's gazelle—25 kg African porcupine—20 kg
10 kg	Honey badger—10 kg
5 kg	Black howler monkey—5 kg
2 kg	Chinese pangolin—2 kg
1 kg	Indian flying fox—1 kg

The largest living organism on Earth, though, is not an animal at all. Dwarfing even the blue whale and the redwood tree, it is a honey fungus in the Blue Mountains of Oregon in the USA. The *Armillaria solidipes* grows underground and extends over an area of 9.6 km². It is somewhere between 1900 and 8650 years old, with a mass that's been estimated as around 500 tons, or five blue whales.

Sink or swim?

What floats and what doesn't? It all depends on the density: specifically whether the average density of the object is more or less than that of water.

Every school kid must have heard the old riddle: 'which weighs more, a ton of feathers or a ton of lead?' The point here is that even though they weigh the same, somehow the lead **feels** as if it should be heavier, and that's because it's so much denser. In any given volume of the denser substance, there is more mass packed in. So we learn about density. Our sense of density may be ill-defined, but it's there. It's a compound measure: the mass of a thing divided by the volume of that thing to give units of mass/volume, for example kg/m^3.

While density can perfectly reasonably be applied to a specific object ('the density of this apple is 0.75 g/cm^3'), it is also often used in reference to a substance. Because it is a compound measure, we can say that the density of, say, pure gold is 19.3 g/cm^3. That's not dependent on what shape that gold is formed into, or how much gold there is in the sample. It's simply the ratio of mass to volume.[101] It's a way of saying how heavy stuff is, rather than how heavy things are.

In the section on mass, we saw that the original definition of a kilogram was the mass of one litre of water at 4 °C (the temperature of greatest density). So that gives us an obvious landmark number to start with.

> **Landmark number** The density of water is 1 kg/dm^3 or equivalently, 1 kg/L, 1 ton/m^3, or 1 g/cm^3.

A comparison with water's density tells us if an object or a substance will float or not. When an item is immersed in water, the weight of water displaced will cause an upward force on the object, and the amount of upward force is the same as the weight of the water that's been displaced.

If the object is denser than the water, for example an iron cannonball, then the upward force is too small to oppose the weight of the object and it sinks.[102] If the object is less dense than water, for example an apple, then the object will sink until just as much water is displaced as would match the weight of the object. So an apple bobs low in the water, while a floating beachball will ride high on the waves.

---

[101] This is rather interesting: it's another conceptual leap, this ability to talk in a semi-abstract way about the qualities of a substance, without having to reference a specific object.

[102] This is called 'Archimedes' Principle'. More about Archimedes later.

An iceberg has a density that is about 90% of that of the seawater in which it floats (which is about 2½% denser than pure water). A floating iceberg must displace sufficient water to equal its own weight, which means it shows only $^1/_{10}$ of its volume above the waterline.

Balsa wood, when freshly felled, is only a little less dense than water and can scarcely float. However, after kiln-drying for two weeks, it reaches a density of just 16% that of water and so floats extremely well. Thor Heyerdahl, trying to demonstrate theories of migration, sailed a balsa wood raft, the *Kon-Tiki*, half-way across the Pacific Ocean. On the other hand, woods such as ebony and lignum vitae are heavier than water, and will sink.

The average density of the human body changes, even as you breathe. Roughly speaking, the human body weighs the same as water. That is to say, without air in your lungs, you would probably sink in both fresh and saltwater. With an average amount of air in your lungs, you will float in saltwater but sink in fresh, and with a great big breath you might float in freshwater. The extreme salinity of the Dead Sea in Israel gives it a density that is 24% greater than that of water. So, in that salty sea, anyone can float with ease.

In good spirits

The strength of alcoholic drinks is nowadays almost always quoted as 'percentage alcohol by volume'. Beer would be typically around 5% ABV and wine between 10% and 15%, while a gin might be between 40% and 50%. But you often hear people talking about a very strong drink as, say, '70% proof'. The pedant in me objects!

The use of the word 'proof' signals a very different way of measuring the strength of alcoholic drinks (in the UK at least), one that's based on the density of the liquid, and not the proportion of alcohol, which means that to quote a 'percentage proof' is quite wrong.[103]

Alcohol is highly valued and so it is highly taxed—and has been for many centuries. Taxation needs measurement, both of quantity and of strength. An early test of the strength of spirits supplied to the British Navy was to mix

---

[103] This would be the old original definition of 'proof', still current in the UK. In current usage in the USA, 'proof' is defined as twice the ABV measurement. What use that is to anyone I don't know, apart from being able to boast of a drink that has a proof measurement greater than 100.

some of the liquid under test with a small quantity of explosive powder to see if it would burn readily or fail to ignite. If it resisted burning, it was 'under proof'. If it burned, it was at proof or overproof. So a benchmark was set.

But the revenue officers wanted finer grades of testing, not just a single underproof/overproof point, and, besides, they wanted to test beer (which would never ignite), and not just spirits. Without the laboratory technology we have today, that meant they would need a way of testing density. Alcohol is lighter than water (at about 79% of the density), and by measuring the density of the drink under test, you can establish a scale that relates the density to the alcohol content. And so, by calibrating density with the ignition test, '100 degrees proof' became defined as the strength of an alcoholic liquid with a density of $^{12}/_{13}$ of the density of water.[104] Testing was done using a hydrometer, which is a calibrated float having a graduated stalk that rises out of the liquid, and which could be weighted down with standardised metal discs. A less dense liquid (which is therefore more alcoholic) will cause the float to sink lower, and so a measurement of density, expressed as degrees proof, can be read off the graduation marks on the stalk, and the right level of tax can be imposed.

Here are the densities of some liquids:

- Alcohol—790 kg/m^3
- Olive oil—800 to 900 kg/m^3
- Crude oil—variable, but around 870 kg/m^3
- Fresh water—1000 kg/m^3
- Seawater—1022 kg/m^3
- Brine—1230 kg/m^3 (close to the Dead Sea salinity)

The Plimsoll line is a marking on the side of a ship that is used for ensuring that the ship is not overloaded. In fact, the marking contains a whole variety of lines, since the buoyancy of the ship will depend on the conditions of the seas it is navigating. The Plimsoll[105] line needs to cater for seawater and freshwater and for a range of temperatures.

---

[104] The calculation behind this is complicated by the fact that when alcohol and water are mixed, some volume is lost (around 4% for a 50:50 mixture)

[105] Named for Samuel Plimsoll, an English politician, whose efforts led to legislation passed in the British Parliament in the 1870s that mandated the use of the Line. Tennis shoes were named 'Plimsolls' based on the supposed resemblance of the band around the shoe to a Plimsoll line on a ship.

Take a look at this list of densities of some other substances. I'm struck by how (relatively) light concrete is, and, generally, how much heavier the metals are than the rocks. Who would have thought the light metal aluminium and the heavy stone granite would be so similar in density?

- Medium concrete—1500 kg/m^3
- Limestone—2500 kg/m^3
- Aluminium—2720 kg/m^3
- Granite—2750 kg/m^3
- Iron—7850 kg/m^3
- Copper—8940 kg/m^3
- Silver—10,490 kg/m^3
- Lead—11,340 kg/m^3
- Gold—19,300 kg/m^3
- Platinum—21,450 kg/m^3

Why did Archimedes leave his bath?

'Eureka!' shouted Archimedes and jumped out of his bath, running naked down the street, 'I've found it!' Or so we are told. What had Archimedes found, and why was this important?

It wasn't just that entering the bath would cause the water to rise and possibly spill. What he realised was that he could use this behaviour to do precise measurements. If he could immerse an object in a vessel of water that was filled to the brim, the amount of water spilled ('displaced') when he dunked the object would be equal to the volume of the object. Since he already had equipment that could weigh things to a reasonable precision, now that he could also measure volumes of irregular objects with complex shapes, he had what he needed to compute the density of those objects.

And this solved a problem he had been set by King Hiero of Syracuse, to tell whether or not a crown was made of pure gold, or whether the goldsmith had replaced some of the gold supplied with silver.

As you see from the list of densities, the density of gold is nearly twice that of silver. So he wouldn't have needed huge accuracy in his calculations: if the goldsmith had swapped out just a quarter of the gold and replaced it with silver, the volume of the crown would be 21% greater than it should be, a discrepancy easily detectable by Archimedes.

What weighs a ton?

1 g to 1 kg

1 g   Mass of a Japanese 1 yen coin—1 g

2 g   Mass of a US penny (1c piece)—2.5 g

5 g   Mass of a US quarter (25c piece)—5.67 g

10 g   Mass of a UK pound coin—9.5 g

20 g   Mass of a house mouse—17 g

50 g   Mass of heaviest allowable golf ball—45.9 g

100 g   Mass of an alkaline D battery—135 g

200 g   Mass of an iPhone 6—170 g

500 g   Mass of an iPad Air—500 g

1 kg   Mass of a medium-sized pineapple—900 g

1 kg to 1000 kg (1 ton)

1 kg   Mass of an average human brain—1.35 kg

2 kg   Mass of a brick—2.9 kg

5 kg   Mass of an adult male Siamese cat—5.9 kg

10 kg   Mass of a large watermelon—10 kg

20 kg   Mass of the heaviest allowable curling stone—20 kg

50 kg   Maximum mass of a flyweight professional boxer—50.8 kg

100 kg   Mass of an ostrich (approximate)—110 kg

200 kg   Mass of a motorbike—200 kg

500 kg   Mass of a mature thoroughbred racehorse—570 kg

1,000 kg   Mass of a Cessna 172 aircraft—998 kg

1000 kg to 1 million kg (1 ton to 1000 tons)

1000 kg   Mass of a loaded MQ-1 Predator military drone—1020 kg

2000 kg   Mass of an adult male walrus—2000 kg

5000 kg   Mass of an adult male African elephant—5350 kg

10,000 kg   Mass of the *Apollo* Excursion Module Moon lander—15,200 kg

20,000 kg   Mass of the *Apollo* Command Module—28,800 kg

50,000 kg   Mass of a Gulfstream G650 business jet—45,400 kg

100,000 kg   Mass of an M1 Abrams tank—62,000 kg

200,000 kg   Mass of a blue whale—190,000 kg

500,000 kg   Mass of the SpaceX Falcon 9 launch rocket—542,000 kg

1 million kg   Mass of the biggest redwood tree—1.2 million kg

More than 1 million kg (more than 1000 tons)

2 million kg   Mass of water in an Olympic swimming pool (minimum depth of 2 m)—2.5 million kg

5 million kg   Mass of space debris in orbit around Earth—5.5 million kg

10 million kg   Mass of the Brooklyn Bridge—13.32 million kg

20 million kg   Mass of annual world silver production (2014)— 26 million kg

50 million kg   Mass of the *Titanic*—52 million kg

100 million kg   Mass of a supercarrier aircraft carrier—64 million kg

200 million kg   Mass of the Empire State Building—331 million kg

500 million kg   Mass of oil capacity of a TI-class supertanker— 518 million kg

1 billion kg   Mass of the Golden Gate Bridge—805 million kg

2 billion kg   Mass of water in the Hoover Dam—2.48 billion kg

5 billion kg   Mass of the Great Pyramid of Giza—5.9 billion kg

10 billion kg   Mass of water in Buttermere (a lake in the English Lake District)—15 billion kg

20 billion kg   Mass of water in Bassenthwaite Lake (in the English Lake District)—28 billion kg

50 billion kg   Mass of water in Haweswater reservoir (in the English Lake District)—85 billion kg

100 billion kg   Mass of water in Rutland Water—124 billion kg

200 billion kg	Mass of water stored in London's storage reservoirs—200 billion kg
500 billion kg	Mass of all the humans on Earth—358 billion kg
1 trillion kg	Mass of all the Earth's land mammals—1.3 trillion kg
2 trillion kg	Mass of annual world steel production (2014)—1.665 trillion kg
5 trillion kg	Mass of annual world crude oil production (2009)—4 trillion kg
10 trillion kg	Mass of annual world coal production (2013)—7.82 trillion kg Mass of comet 67P/Churyumov–Gerasimenko (Rosetta's comet)—10 trillion kg
20 trillion kg	Mass of water in the Fort Peck Dam (USA)—23 trillion kg
100 trillion kg	Mass of water in Lake Geneva—89 trillion kg
200 trillion kg	Mass of Halley's Comet—220 trillion kg
500 trillion kg	Mass of all the Earth's biomass—560 trillion kg
1 quadrillion kg	Mass of carbon stored in the Earth's atmosphere—720 trillion kg
2 quadrillion kg	Mass of Deimos (moon of Mars)—2 quadrillion kg
5 quadrillion kg	Mass of carbon stored in the world's coal deposits—3.2 quadrillion kg
10 quadrillion kg	Mass of Phobos (moon of Mars)—10.8 quadrillion kg
20 quadrillion kg	Mass of water in the Great Lakes of North America—22.7 quadrillion kg

Heavy, man, heavy

Mass of the **Earth** (5.97 septillion kg) is
  40 × mass of **Ganymede** (moon of Jupiter) (148.2 sextillion kg)

Mass of an adult male **giraffe** (2500 kg) is
  40 × mass of the average **human** (62 kg)

Mass of a **Gulfstream G650** business jet (45,400 kg) is
  100 × mass of a **concert grand piano** (450 kg)

Mass of a **rhino** (2,300 kg) is
  4 × mass of a thoroughbred **racehorse** (570 kg)

Mass of a **double bass** (10 kg) is
  25 × mass of a **violin** (400 g)

Annual world **steel production** (2014) (1.665 trillion kg) is
  5000 × mass of the **Empire State Building** (331 million kg)

Mass of a **brick** (2.9 kg) is
  500 × mass of a **Bic pen** (5.8 g)

Mass of oil capacity of a **TI-class supertanker** (518 million kg) is
  10 × mass of the *Titanic* (52 million kg)

Getting Up to Speed

Putting a Value on Velocity

Which of these is the fastest?
- ☐ Top speed attained by a human-powered aircraft
- ☐ Top speed of a giraffe
- ☐ Top speed attained by a human-powered watercraft
- ☐ Top speed of a Great White shark

Blue Riband

On 15 July 1952, the passenger liner *SS United States* reached Ambrose Light, a lightship stationed in Lower New York Bay. The liner had just crossed the Atlantic in 3 days, 12 hours, and 12 minutes, at an average speed of 34.51 knots (almost 64 km/h) and had thereby claimed the Blue Riband, a distinction unofficially awarded to the commercial passenger vessel making the westward crossing of the North Atlantic at the highest rate of knots. It had beaten the achievement of the *RMS Queen Mary*, which had held the prize for 14 years, and it was the last vessel to win the Blue Riband according to this strict definition.

Over the previous 115-odd years, the Blue Riband had been a badge of honour for the shipping lines competing for business on that lucrative route, and their ships had become renowned for luxury. During that period, advances in technology had propelled the ships faster and faster, and crossing times had fallen from around two weeks to half a week as speeds had increased from around 8½ knots (approximately 16 km/h) to over 30 knots.

But the very commercial pressures that drove the rivalries were bringing the competition to a natural end. In 1927, Charles Lindbergh had flown across the Atlantic, and by 1938 the first commercial plane had made the crossing. In 1939, Pan American Airways launched a regular service between New York and Marseilles in France, and later in the same year to Southampton in the UK. The time taken for these early flights was around 30 hours.

After the Second World War, in 1947, Pan American instituted a regular service between New York and London, effectively signalling the end of the Blue Riband competition. Even three and a half days was too long when those with enough cash could fly across the Atlantic in half a day. The commercial imperative for speed had given the prize to the airliners. People still talk loosely of present-day Blue Riband holders, even if the strict definition is not adhered to. The current record for the fastest crossing of the North Atlantic by a commercial vessel is held by the *Fjord Cat* (formerly *Cat-Link V*), which made an eastbound crossing in 1998, taking 2 days, 20 hours and 9 minutes, at an average speed of 41.3 knots (76.5 km/h).

> **Landmark number** Top speed of a passenger liner—64 km/h

Measuring speed

We're impressed by speed. Whether it is Usain Bolt on the track at the Olympics, the anticipated challenge to the World Land Speed record by the Bloodhound SSC project, or the speed of Andy Murray's serve, we see ever-greater speed as ever-greater achievement. More than that: we say 'Time is money', and that puts a value on speed. Indeed, as the competition for the Blue Riband contest shows, for much of the twentieth century, speed was almost synonymous with progress.

Speed is a compound measure. We take a spatial measurement—distance travelled—and divide by a time measurement—the time taken to travel the distance. Any measure when divided by time becomes a rate, and can be thought of as a speed. So a printer's speed may be measured in terms of pages per minute, a data transmission speed in bits per second, and even speed of drumming in strokes per minute, as recorded in the WFD (World's Fastest

Drummer[106]) competition. In this chapter, though, we'll concentrate on the fundamental meaning: distance over time.

The units we use for speeds usually follow the pattern of 'unit of distance per unit of time'. But not always: the Blue Riband speed records above were quoted in knots, as is conventional for ships. Why? In past centuries, the speed at which a ship was moving (relative to the water) was determined by throwing overboard a wooden panel (a 'log') attached to a line, which was allowed to spool out as the ship left the log behind, 'stationary' in the water. The line had knots tied in it, spaced eight fathoms apart, and as the line ran out over 30 seconds, the sailor counted how many knots went past. The count of knots was the speed of the vessel, and was measured, naturally, in 'knots'. The spacing of the knots and the time taken for counting formed a calibration that made one knot equal to one nautical mile[107] per hour.[108] Measuring speed in this way makes it clear that it involves establishing a ratio between two kinds of counting: the counting of the knots and the counting of the seconds.

Going like the wind

If you listen to the BBC's Radio 4 service, you will hear the Shipping Forecast. In the main, this consists of an announcer reading out what amounts to a database of information related to a collection of 31 named sea areas around Britain, from 'Viking' in the far Northeast to 'Southeast Iceland' in the far Northwest. A typical excerpt might be something like this:

Viking North Utsire South Utsire Southeast 4 or 5 increasing 6 or 7 veering South 4 or 5 later. Occasional rain. Good with fog patches becoming moderate or poor.

---

[106] Current records: 1208 strokes per minute by hand, 1034 strokes per minute for feet.

[107] A nautical mile is based on lines of latitude. If you sail directly North or South for a distance equivalent to one degree of latitude, you have covered 60 nautical miles. So a nautical mile is $^1/_{60}$ of $^1/_{90}$ of the distance from pole to equator. That's $^1/_{5400}$, whereas the first definition of the kilometre was one ten-thousandth of the same distance. This means that a nautical mile is somewhat less than 2 km.

[108] And a record was kept of these speeds. They were, of course, written down in the ship's log book.

Each element in the Shipping Forecast lists the sea areas to which it applies (in this case, Viking, North Utsire, and South Utsire), the forecast wind conditions, precipitation, if any, and finally visibility conditions. The wind conditions give the direction (in this case, initially 'Southeast'), and the forecast wind speed ('4 or 5'). But how strong a wind is '4 or 5, increasing 6 or 7'?

The wind numbers are from a scale devised and introduced by Francis Beaufort in 1805 for the purposes of regularising wind speed measures, and were first used on the voyage of *HMS Beagle*, the same trip on which Charles Darwin was a passenger. The scale is notable in that it was based on observation of natural phenomena, and did not require explicit measurement— although there are now specific wind speeds associated with each scale point.

Here's a summary of the scale:

Force and classification	Sea observation	Speed	
		(knots)	(km/h)
0, Calm	Sea like a mirror	<1	<1
1, Light air	Ripples, but no foam	1–3	1–5
2, Light breeze	Small wavelets, do not break	4–6	6-11
3, Gentle breeze	Large wavelets, some whitecaps	7–10	12–19
4, Moderate breeze	Small waves with breaking crests	11–16	20–28
5, Fresh breeze	Moderate waves, some spray	17–21	29–38
6, Strong breeze	Long waves, white foam crests, spray	22–27	39–49
7, High wind	Sea heaps up, streaks of foam	28–33	50–61
8, Gale	Moderately high waves, spindrift	34–40	62–74
9, Severe gale	High waves, crests rolling over. Spray reduces visibility	41–47	75–88
10, Storm	Very high waves, tumbling. Large patches of foam make the sea white	48–55	89–102
11, Violent storm	Exceptionally high waves. Very large patches of foam, covering much of the surface	56–63	103–117
12, Hurricane	Huge waves. Sea completely white, air filled with spray	64+	118+

Landmark numbers
Gale—roughly 60 km/h
Storm—roughly 90 km/h
Hurricane—roughly 120 km/h

A hurricane is defined by wind speeds that are in excess of 64 knots (118 km/h), but there is a further level of classification for hurricanes themselves. In 1971, Herbert Saffir and Robert Simpson developed a scale based on wind speed, to assign categories to hurricanes:

Category	Observation	Speed	
		(knots)	(km/h)
1	Will produce some damage. Roof tiles blown off	64–82	118–153
2	Will cause extensive damage	83–95	154–177
3	Devastating damage will occur. Structural damage to small residences	96–112	178–208
4	Catastrophic damage will occur. Structural failure on small residences	113–136	209–251
5	Catastrophic damage will occur. Few structures survive intact	137+	252+

Speed limits

When railways were introduced in Britain, it was feared that the great speeds involved would endanger human health: 50 miles per hour was thought to be more than the body could withstand. In a sense, they were right: the faster one travels, the greater the risk in the event of an accident. That's why speed limits are in place on roads and on rail. But some speed limits are less arbitrary—those set by nature.

Sound is the propagation of vibrations through a medium, typically air, and the speed at which those vibrations are transmitted depends on the nature of the substance. And this is the basis of the definition of the 'Mach' system of measuring speed. A Mach number is a speed measurement that has no units: it is simply the ratio of the speed of the vehicle (usually a plane, but possibly a 'car' attempting a land speed record) to the speed of sound in the local environment.

The speed of sound is nominally quoted as 1236 km/h, but it varies with temperature and altitude. Whatever the speed of sound may be in local conditions, it can always be called Mach 1.

But there is one speed limit that is, as far as we know, an absolute. That is the speed of light.[109] Einstein showed that as the speed of an object approaches the speed of light, so further acceleration of that object becomes more and more difficult. No matter how much additional force is provided, nothing can accelerate an object beyond the speed of light. The speed of light can be regarded as the natural unit for measurement of speed in the sense that it emerges naturally from physical constants and requires no assumptions or arbitrary definitions.

> **Landmark numbers**
> * Speed of sound—1236 km/h
> * Speed of light—1.08 billion km/h or 300 million m/s

Blue Birds, Bluebirds, and Bloodhounds

In 1924, Malcolm Campbell drove a Sunbeam car over Pendine Sands on the shores of Carmarthen Bay on the south coast of Wales at over 235 km/h, thereby breaking the world land speed record. He went on to break the record repeatedly, on land and on water, in a succession of vehicles and vessels, each of which was called *Blue Bird*. His final record-breaking run was over the Bonneville salt flats in Utah in 1935, when he reached a speed of almost 485 km/h.

His son, Donald Campbell, continued the family tradition, capturing the world land speed record in July 1964 by reaching a speed of 648.73 km/h in a car called *Bluebird CN7*.[110] On 31 December of that same year, he broke the world water speed record, in *Bluebird K7*, with a speed of 444.71 km/h.

But on 4 January 1967, Donald Campbell died in an attempt to push *Bluebird K7* to a water speed of 480 km/h, on Coniston Water in the English Lake District. The craft and Campbell's body were recovered many years later, between October 2000 and May 2001.

At the time of writing, the World Land Speed record is held by a vehicle called the *ThrustSSC*, and was set in October 1997, reaching a speed of 1228 km/h over

---

[109] We should really say the speed of light in a vacuum. As with the speed of sound, the speed of light varies with the medium through which it travels, but is never greater than when in a vacuum.

[110] Malcolm Campbell's machines were *Blue Birds*, Donald Campbell's were *Bluebirds*.

a distance of 1 mile. It was the first time that the speed of sound had been broken on land. However, the holder of that record, Andy Green, is now preparing for a new attempt at the record with the *Bloodhound SSC*, a jet- and rocket-powered vehicle that aims to reach 1690 km/h. Its attempt on the record is planned for October 2018, on the Hakskeen Pan in Northern Cape, South Africa.

> **Landmark number** World Land Speed record—1228 km/h

Terminal velocity

When you drop an object, it accelerates due to the action of gravity, speeding up as it falls. But there's an opposing force too: air resistance. And while the force of gravity stays more or less constant for any particular object (it's proportional to the object's mass), the air resistance is proportional not only to the surface area of the object, but also to the square of its speed through the air. This means that as it speeds up, the downward force stays the same, while the upward force increases, reducing the acceleration. In time, the falling object will approach a limiting speed where the upward and downward forces balance each other, and the falling object will no longer accelerate. This is the object's terminal velocity and it depends on the density of the air and the cross-sectional area that the object presents to the air.

Urban legend has it that a penny dropped from the Empire State Building would kill a person. Freak injuries excepted, this just isn't true. The terminal velocity of a penny falling from a height is a little over 100 km/h and at that speed, a penny will cause bruising, but not serious damage.

A skydiver in the spread-eagled freefall posture will reach a terminal velocity of around 200 km/h, which is approached after around 12 seconds of fall. If the skydiver then pulls in their arms and legs, they reduce the area exposed to air resistance, and the terminal velocity can increase to around 300 km/h. And if every effort is made to reduce drag from the air, speeds of over 500 km/h can be reached. Finally, once the parachute is properly deployed, the huge surface area brings the terminal velocity right down to something like 20 km/h, for a safe landing.

> **Landmark numbers**
> - Terminal velocity for a penny piece—100 km/h
> - Terminal velocity for a skydiver—200 km/h

Escape velocity

Escape velocity is the upward speed that a projectile needs to reach in order to escape the gravity of the Earth (or another planet, moon, or celestial body). As the projectile travels further from the centre of mass, so the gravitational force pulling it back will reduce. If it starts off at a speed that is anything less than escape velocity, eventually gravity will ensure that it falls back: any speed above escape velocity, and gravity has lost the battle.

Escape velocity from the surface of the Earth is around 11.2 km/s, or just over 40,000 km/h.[111] To escape the Moon's weaker gravity would require a velocity of around 8600 km/h.

Landmark number Escape velocity from Earth—40,000 km/h

Orbital velocity

The rockets that deliver satellites into Earth's orbit and crew to the International Space Station have two criteria to meet: they must get high enough to reach the altitude of the desired orbit and, once they reach that height, they must be moving fast enough to maintain the orbit.

In a so-called low Earth orbit of 200 km (half as high as the ISS), an orbital velocity of a little over 28,000 km/h is needed.[112] A geostationary orbit, such as television satellites use, is at around 36,000 km high, and needs a slower orbital velocity of 11,160 km/h.

Landmark number Speed of a TV satellite—11,160 km/h

Faster and faster

Here's a list of speeds from slow to fast:

- A snail's pace is around 10 m/h, so 0.01 km/h.
- Slow but steady, a tortoise typically moves at about ½ km/h, but can reach 1 km/h.

---

[111] This number seem familiar? The Earth's equator is approximately 40,000 km—so escape velocity is equivalent to a speed that would circle the Earth in an hour.

[112] The ISS itself, at 400 km high, has an orbital velocity of 27,600 km/h.

- Planning a day's hiking? Anticipate a comfortable walking speed on the flat of 5 km/h.
- The fastest person on foot (in 2016), Usain Bolt reached 44.7 km/h between the 60th and the 80th metre of a 100 m race in Berlin in 2009.
- A horse can gallop at 45 km/h or a little more.[113]
- Planning a road trip? Reckon on driving 100 km/h on good roads and conditions.
- Prefer the train? Japan's Bullet Train travels at 320 km/h, as does France's TGV (*train à grande vitesse*) (and it's reached 570 km/h in test conditions)
- You'd rather fly? A flight on a commercial airliner will carry you at 850 km/h.
- The fastest speed ever achieved on land—at the time of writing— 1228 km/h by the *ThrustSSC* vehicle.
- A challenge to this record is anticipated in 2018 by the *Bloodhound* project, which aims to smash the record by travelling 33% faster.
- Even when we're standing still, the Earth is rotating about its axis and carrying us with it at 1675 km/h (luckily the air around us is also travelling at that same speed).
- If you snagged a flight on *Concorde* when it was operating, your speed would have reached 2140 km/h.
- The fastest military/experimental flight clocked in at 3500 km/h.
- Superman supposedly flies faster than a speeding bullet. That means he flies faster than a round moving at around 5000 km/h.
- To enter Earth's orbit, a rocket must reach 7.9 km/s = 28,440 km/h.
- Every year, the Earth covers about a billion kilometres around the Sun, and that means it moves relative to the Sun at a speed of 30 km/s = 107,000 km/h.
- The Sun is moving within the Milky Way galaxy at around 70,000 km/h, relative to its local neighbourhood of stars.
- The Sun's local neighbourhood itself rotates around the centre of the galaxy over the course of a galactic year of around 225 million Earth years. This means it's moving through space at 792,000 km/h.

---

[113] The famous 'Pony Express' of the American West operated for only 19 months, but could get a message across the North American continent, a distance of over 3200 km, in 10 days. The coming of the telegraph made it redundant. Again, commercial demand for speed was the spur to, and measure of, progress.

- And the fastest speed there is? Of course, this is the universe's ultimate speed limit, the speed of light, at around 1 billion km/h.

Warp speed and beyond

As far as we know, nothing can travel faster than light. If speeds faster than light were observed, this would be a violation of Einstein's Special Theory of Relativity, and that would shake the foundations of modern physics. (Not that this would be a bad thing—science thrives on new results that require explanation.) Even when scientists working on the OPERA project recorded what appeared to be neutrinos moving faster than light between the CERN laboratory on the French–Swiss border and the LNGS laboratory at Gran Sasso in Italy, 731 km away, they were sceptical of their own results and published them in the spirit of 'help us find out how we messed up'. They had indeed messed up, and the anomaly was fully explained by experimental errors.

So, for the time being, Science Fiction is the only realm where faster-than-light travel happens. Speaking of that, how fast are *Star Trek* Warp speeds, anyway? The collective *Star Trek* fan base have collated all available evidence, and the results run a little like this:

Warp factor 1 is the speed of light. On this, all are agreed. From this point on, however, there is a divergence between the warp factor–speed relationship used in The Original Series (TOS) and the accompanying movies, and those used in the series starting with The Next Generation (TNG). In the period in which TOS is set, Warp factor 2 is equivalent to 8 times the speed of light (c), Warp factor 3 is 27 times c, and so on, the key factor being that you take the cube of the Warp factor to calculate how much faster than light the *Enterprise* is travelling. In the later series, some imagined technology upgrade has boosted these speeds. Now Warp factor 2 is 10 times c, Warp 3 is 39 times c, and Warp factor 10 takes your speed to infinity. Now you know.

Moving swiftly on...

Fastest speed attained by a **propeller aircraft** (870 km/h) is
   25 × the top speed of a **bottlenose dolphin** (35 km/h)

Top speed of a **horse** (88 km/h) is
   2 × the fastest speed by a **human-powered aircraft** (44.3 km/h)

Top speed of a **cheetah** (120 km/h) is
   2.5 × the top speed of a **domestic cat** (48 km/h)

Fastest speed of a **human powered vehicle** (144 km/h) is
   2 × the top speed of an **African wild dog** (72.5 km/h)

Top speed of a **lion** (80 km/h) is
   2 × the top speed of a **great white shark** (40 km/h)

Fastest speed of a **wind-powered watercraft** (121.2 km/h) is
   2.5 × the top speed of an **orca (killer whale)** (48.3 km/h)

Fastest speed attained by a **jet aircraft** (3530 km/h) is
   4 × the speed of a **commercial airliner** (880 km/h)

Speed of the **Earth's rotation around the Sun** (107,000 km/h) is
   50 × the speed of *Concorde* passenger SST (2,140 km/h)

INTERMISSION

Time to Review and Reflect

So many people today—and even professional scientists—seem to me like someone who has seen thousands of trees but has never seen a forest.   **Albert Einstein**

The Philip Glass opera *Einstein on the Beach* is filled with numbers. In several of the sections, the voice parts consist simply of the recitation of sequences of numbers. The opera's libretto starts with these words:

> One, two, three, four,
> One, two, three, four, five, six,
> One, two, three, four, five, six, seven, eight.
> One, two, three, four,
> One, two, three, four, five, six,
> One, two, three, four, five, six, seven, eight.
> One, two, three, four,
> One, two, three, four, five, six,
> —, two, three, four, five, six, seven, eight.

This first section is listed as *Knee Play 1*, and there are four other *Knee Plays* through the work, short pieces framing and linking the main acts. They serve a practical role (for scene changes) and as articulation points, joints between the substantial elements.

Think of this short section of the book as a kind of 'Knee Play'. This joint is in two parts, closely coupled. The first part reflects on the variability and distribution of the numbers that we actually come across in this practical day-to-day world. The second part of the knee joint connects with those reflections and describes the last of the five techniques for grappling with big numbers, in preparation for talking about numbers that go beyond the confines of this world.

Numbers in the Wild

Variability and Distribution

Number spotting

Here's a game I played in my head as a kid growing up in South Africa, bored with sitting in a car on long road trips. Look at the registration plates of the cars around you,[1] and wait until you spot one whose number starts with the digit 1. Having bagged that one, spot a registration starting with 2. Then look for a 3, then a 4, and so on. Of course, once you have 9, you move to two digits, so you next have to look for a number starting with 10, then 11, and so on. As numbers get bigger, so the game gets harder, and so the satisfaction of spotting your next number gets greater. I can't quite remember the highest number I ever spotted, but it was in the hundreds.

Now it's obvious that spotting 10 is going to be harder than spotting 1, since there are two digits to match, but what is more surprising is that spotting 9 is also a lot more difficult than spotting 1. And that's because the local registration authorities allocated the numbers sequentially in each location. In no location could there be a 9 registration without there being a 1 registration first, and you could never have a two-digit number starting with '9' without there being at least eleven numbers starting with 1 (1, and 10 through to 19). So 9s could never outnumber 1s, and in the great majority of cases, 1s would outnumber 9s. In the same way, 9s could never outnumber 5s, nor could 5s outnumber 1s—the natural distribution of first digits is always going to be one that favours smallness.

---

[1] At that time, South African registration plates consisted of an alphabetic group, designating a local authority, followed by a number of typically 1 to 4 digits (my parents' car was CAP 560). This game, sadly, cannot be played in the UK where I now live—the format of the registration numbers is just wrong.

Benford's law

I didn't know then, but I do now, that this is a manifestation of Benford's law, which describes the distribution of leading digits of numbers. This law can quantify this tendency and describe it mathematically. A set of numbers that follows Benford's law will have around 30% of the numbers starting with '1' as the leading digit, and only around 4% of the numbers starting with '9'. What is truly startling is that very many sets of numbers gathered from real-life contexts actually do satisfy this law.

The law is sufficiently robust that it is used as a test for fraudulent accounting. Unless the fraudster is savvy, and knows about Benford's law, then when, for example, inventing a number to cook the books, he would be more likely to create a number starting with, say, 7, than would occur by chance. Benford's law says only 5.8% of numbers start with 7. So if we analyse a company's books, or a country's reported election statistics, and find a proportion of 7s significantly more than 5.8%, we should start smelling fraud.

Benford's law put to the test

IsThatABigNumber.com has a collection of numbers gathered purely for the purpose of putting other numbers into context. The numbers have been chosen not for any quality of the numbers themselves, but because they represent good comparators—they may relate to everyday standardised items like the length of a baseball bat, or the height of well-known buildings or structures such as the Eiffel Tower, or to population counts, such as the number of elephants in the world. It's as disparate a collection of quantities as you could wish to find: a collection of numbers from the wild.

I put these numbers to the Benford's test and here are the results:

Leading digit	Benford says	I found
1	30%	28%
2	18%	16%
3	12%	13%
4	10%	11%

Leading digit	Benford says	I found
5	8%	9%
6	7%	7%
7	6%	7%
8	5%	5%
9	4%	4%

It's not an exact fit, but it is astonishing to see such regularity emerge from data that has been collected from such divergent sources and mixed together without any sort of categorisation.

There's something subtle going on here. When a mathematician contemplates numbers, it's as if she is studying them in laboratory conditions. She can construct and study any numbers she chooses to create, and they will be perfect specimens. For example, we know that (by definition) the counting numbers go on forever without gaps. And we understand that the real numbers between 900 and 901 are just as closely packed as those between 100 and 101. And π (pi) is an exact number even if its decimal expansion never ends.

But what if, instead of thinking about numbers that come from a Platonic realm of ideal forms, where all numbers are possible and perfect, we look at the numbers that we really do encounter in life when we examine actual counts and measurements? Well, we find out that the bigger the numbers get, the less frequently they are encountered. It is likely, for example, that the number 728,167,198,612,003 has never before been used in a written text, and it is likely that the number that is one greater than that never will be. So, in a sense, numbers in the wild 'thin out' as they get bigger. It's an odd thought, and a slightly disturbing one, but one we'll return to, and it's important in how we think about big numbers.

Benford's law does **not** apply where the numbers come from a narrow distribution of similar-sized numbers. Take the first digit of a sample of people's heights, and you'll find that, if you're measuring in metres, the vast majority of leading digits will be 1. If you're measuring in feet, expect a very few 4s, mostly 5s and 6s and a very few 7s. But where the numbers show a range of 100 times or more between the largest and smallest, the law seems to apply quite generally.

Distribution of magnitude between 1 and 1000

This is the way that numbers are stored in the IsThatABigNumber.com database: there are two parts to each number, a **significand,** which is always scaled to be in the range between 1 and 1000, with typically three significant figures, and a **scaling multiple**, which signifies a power of 1000: one, a thousand, a million, a billion, and so on up. So, ignoring the multiple for now, for every number in the database there is a significand lying between 1 and 1000. I decided to look at the distribution of these numbers.

It's rather like a version of Benford's law, but instead of looking at the first digit in a base-10 notation, we're looking at the first digit in a base-1000 system. I sorted the significands so that the sequence of numbers starts with 1 (from '1000 years since the building of Great Zimbabwe'), and ends with 998 (from '998 kg is the mass of a Cessna 172'). After I sorted the numbers, I graphed them.

Here's the result: the graph shows the distribution of 2000+ of the numbers in the database, normalised so that each lies between 1 and 1000, and ignoring any scaling factor used to achieve the normalisation:

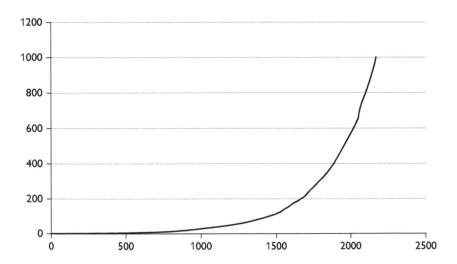

Now, if all the numbers were evenly distributed, we'd see a straight line ascending with constant slope. This, as you see, is by no means an even distribution: in fact, 33.7% of the significands were between 1 and 10, 32.6% of them were between 10 and 100, and 33.7% were between 100 and 1000.

Think about what that means. In our hotchpotch collection of numbers, there are about as many of them that are in the 1–10 range as there are in the 10–100 range, and roughly the same number again that are in the 100–1000 range. If we partition the range of the numbers (1–1000) into bands that increase by the same proportion (each being ten times the size of the previous one), then an equal share of numbers goes into each band.[2]

The graph itself looks pretty smooth, and an experienced eye might lead one to guess from its shape that it is roughly exponential, and that in turn suggests that using a logarithmic scale on the vertical axis could be a good idea.

Logarithmic scale? Don't be scared off. Understanding log scales is a superpower you're about to acquire. For now, just watch what happens when we plot that same graph, but adjust the vertical axis to use a log scale (which will mean that the 1–10, 10–100, and 100–1000 bands are treated as equally sized):

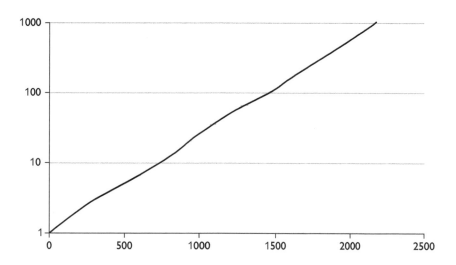

Bingo! That looks like a slightly wobbly straight line. See how our three bands become visible? The straightness of the line shows much more clearly how the numbers are distributed: roughly a third of the numbers are less than 10, roughly another third take us to 100, and the final third take us to 1000. That's

---

[2] In fact, if we split the numbers into bands based not on a factor of 10, but on any other multiple, the same general principle holds: the band 1–4 holds 18.5% of the numbers, the band 4–16 holds 21.9% of the numbers, 16–64 holds 18.3%, 64–256 holds 21.9%, while the slightly undersized band 256–1000 holds 14.4%. Each band holds about a fifth. That's a clue that we're dealing with an exponential effect here.

an insight that is hard to pick up visually without switching to a log scale, and it's a little taster for the next part of the book.

Back to the main topic. In some ways, this result, that the numbers in a very arbitrary collection of numbers 'thin out' as we go from 1 to 1000, in such an orderly way, seems quite astonishing. On the other hand, it confirms a very natural intuition, that of all the numbers that surround us, there really are more of them that are small numbers, and those that are big are rarer. This might go some way to explaining why we are less comfortable around big numbers—we simply come across them less often.

Stand up and stretch your legs

If you've stuck with me this far, thank you. I hope you've enjoyed the ride. We've seen some big numbers, and have explored some techniques for understanding them. But things are about to get a little wilder as we head into the next chapters, and we start looking at numbers that go far beyond our day-to-day experience. To take that leap, we need one more technique for understanding big numbers under our belt, and it's a biggie. To prepare you for that, let us recall and link together some of the points we've already covered:

- Our innate 'approximation' number sense is a **proportionate** sense. The accuracy of our judgement of numerosity stays at roughly the same proportion, around 20% accuracy.
- The words we use for numbers bigger than a thousand form a list based on multiples of a thousand: thousand, million, billion, etc. Beyond a thousand, each new term is a **multiplication** by 1000.
- You've seen ladders of landmark numbers covering distances, time-spans, masses, and more, They've been based on a sequence that I call the money numbers: 1, 2, 5, 10, 20, 50, 100, and so on, following the same pattern as is frequently used for currency denominations. Every step we **multiply** by 2 or 2½. Every 3 steps gives us a multiplication by 10.
- Benford's law tells us that when it comes to numbers that actually are observed in the wild, the numbers thin out. Numbers that start with '9' are fewer than those that start with '1'. And if we look at more than just the first digit, we see that they thin out on a **proportionate** basis.

All of these use, or suggest, comparisons based on proportions and not differences. We make kindergarten numbers bigger by adding, and we compare them by subtracting.[3] A school ruler allocates the same length of edge to each successive number. That's a **linear** scale, and it suits adding and subtracting.

But when we talk about much bigger numbers, and numbers that grow rapidly, then more and more we become concerned with **proportions** and **ratios**. For measuring bigger and bigger numbers, it might be useful if we had a ruler that, say, allocated the same length of ruler to each **proportionate** increase in scale.

Whisper it. We need logarithms. Second to calculus, and perhaps alongside trigonometry, few subjects in school mathematics have as fearsome a reputation as logs. You might have thought that raising to powers—exponentiation—was bad enough, but once you had struggled to master that skill, logarithms asked you to do that process in reverse, and for some people that was taking it a step too far.

That's a huge pity, because when it comes to understanding big numbers, using log scales is a superpower, better than seven-league boots, because each step you take is even bigger than the last. It's a pity, too, because, believe it or not, log scales are a natural way of seeing the world. An example: in the original prize structure for the television game/quiz show 'Who Wants to be a Millionaire', the amount you could win for correctly answering each successive question increased through the show in a way that is very nearly a perfect doubling sequence:

100—200—300—500—1000—2000—4000—8000—16,000—
32,000—64,000—125,000—250,000—500,000—1,000,000.

That's a log-scale ladder with 15 rungs, where each new rung means a doubling of the prize.

I'm hoping I can show you a new way of looking at the world, through the lenses of logarithms, and that this will help you see the very big numbers a lot more clearly.

So don't panic! This is still not a book about maths: it's a book about numbers. There won't be any algebra, and you won't need to calculate any logs. But in the next chapter I will show you how the numerate citizen can put on those log lenses to unleash the superpower of log scales, and to conquer a whole new level of big numbers.

---

[3] After all, that's what 'difference' means.

The Fifth Technique: Log Scales

Comparing the Very Small with the Very Big

What if we measured things using a scale where every step of equal size meant not **adding** a constant amount to the previous number, but **multiplying** by a constant amount? A scale where the step from 'one' to 'ten' was the same size as the step from 'ten' to 'a hundred', and then again, just the same-sized step took you to 'a thousand'. That's a log scale.

Breaking the number line

Suppose I challenged you to represent the following numbers, for comparison with one another, on a single number line:

- The typical height of an African elephant—4.2 m
- The height of the world's first skyscraper—42 m
- The height of the Empire State Building—381 m
- The depth of the deepest mine—3.9 km
- The highest freefall parachute jump—39 km
- The altitude of the orbit of the International Space Station—400 km
- The diameter of Earth's Moon—3480 km
- The altitude of a geosynchronous satellite's orbit around the Earth—35,800 km
- The distance from the Earth to the Moon—384,000 km

That would be a struggle, wouldn't it? If we drew a 30 cm long number line, with the distance from the Earth to the Moon placed somewhere towards the end, then the second largest number, the one about the satellite's orbit, would have to be near the 3 cm point. The diameter of the Moon would have to be at around 0.3 cm, and the space station's orbit would be just about one-third of a millimetre from the origin.

It just wouldn't work. This number line would show a couple of things very clearly: that these steps, these tenfold jumps, are very large, and that the early numbers are very small in relation to the bigger ones. Those are things worth showing, because it emphasises the sense of scale. But this number line loses all representation of the smaller numbers, and is unable to make any more subtle comparisons between the numbers.

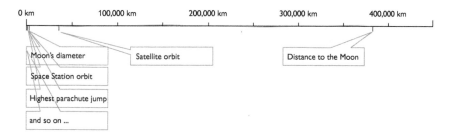

But if we could systematically vary the scale of our number line, if we could allocate equal space for each power of 10, then in nine steps we could go from 1 m to 1 million kilometres (10^9 m), and each of those nine steps could accommodate one of our numbers.

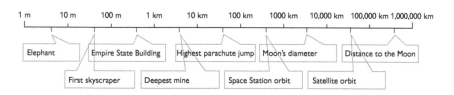

By contrast, this second representation lacks the wow factor of 'this is **so** much bigger than that', but it does show you all the numbers and it also shows, for example, that the spacing between these numbers is actually pretty regular (when you're using the **proportionate ratio** between them). That is, each number is roughly 10 times the previous one.

And that's the power of the log scale, to allow us to make meaningful comparisons between numbers, even when we're dealing with very different scales of measurement.

Moore's law

In the 1960s, Gordon Moore of the computer chip company Intel noticed that the number of transistors that could be accommodated on a microprocessor was increasing exponentially. You may be familiar with this as 'Moore's law' (and, remarkably, this law has held true, more or less, for 50 years, now[4]).

The following example uses a series of numbers that relate to Intel processors from 1972 to 2002, and how many transistors each has. The numbers start at around 2000, and end up at 410 million, a ratio from start to finish of 1:200,000. It's very hard for anyone to easily compare numbers that are so different in scale.

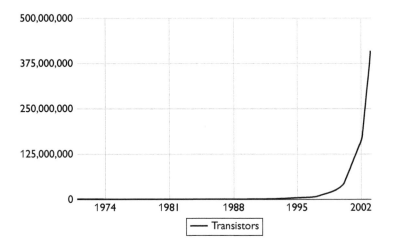

The graph of these numbers looks a bit like a cliff face: it's not possible to see any detail at all of growth in the early years, but then the slope takes off and zooms upwards. It's hard to make any judgement about the nature of the growth that is represented other than this: that the numbers start relatively small, and then get bigger very quickly. In short, we learn little from this graph.

---

[4] This is not a law of nature, of course, just an observation, and it is likely to break down within the next 10–20 years. (But they've been saying that for the last three decades, so I may be proved wrong.)

But now let's look at this through log lenses.[5] Instead of charting the transistors per chip, we'll use the **log** of that number:[6]

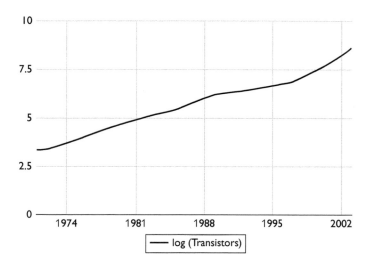

So here we see the power that the log-lenses give us: the ability to compare numbers that are at vastly different scales. Instead of plotting 2250 on the graph, we can plot the log of that number, which is just a bit more than 3, and instead of 410 million, we can use 'about 8½'. Hey presto! We've made the incomparable comparable.

More than that, we can now see some interesting features in this graph. There is, overall, a relatively even slope (slackening a bit in the early 1990s, then picking up again). An even slope on a log graph means a relatively even rate of growth, which is the heart of Moore's law. In fact, Moore's law is precisely equivalent to saying that if we chart the log of the number of transistors on a chip over time,

[5] This is just my way of saying: let's apply a transformation to the thing we are measuring—so instead of showing the number itself, we'll use a scale based on the log of the number.

[6] Don't worry too much about what base we're using for logs: the superpowers of the log scale work whatever base you use. Logs to base 10 are sometimes easier to work with because they are connected to how we write numbers in the base-10 system.

An easy way to mentally 'guess' logs to base 10 is: a bit less than the number of digits that are needed to write down that number. Numbers in the range of 1000 to 9999 need 4 digits and the logs of those numbers range from 3 to just under 4; in the our transistors example above, the figure for 2002 is 410000000 (9 digits), and the log is approximately 8.6.

Because computers use binary numbers (base 2), computer scientists like to use base-2 logs. A base-2 log is closely related to how many bits (binary digits) are used to store a number.

we'll get a straight line. And how steep or flat that line is will tell us what the rate of growth is.

Any process that grows by the same **factor** every time period is showing exponential growth. Such a process will have a constant doubling time, and this is another way of expressing the growth rate.

Moore's law is usually quoted as saying that the number of transistors doubles every year and a half. From the numbers I've used here, we can see that over the 30 years of our graph, the number of transistors has doubled roughly 17 times, and that means a doubling more or less once every 1.7 years, which more or less validates Moore's law over that period.

> Every step you take on a log scale doesn't just **add** a bit more, it **multiplies** a bit more, and that is what gives the log scale its power. On the log scale, joining two distances has the effect of multiplying the numbers they represent, not adding them. And that is why you can travel so far, so quickly, on a log scale.

It's not just for transistors that this is useful: there are plenty of other fields where the numbers involved would break a normal linear scale (each step adds the same amount), and where a log scale (each step multiplies by the same amount) works better. One example is the measurement of earthquakes.

The Richter scale

A seismograph measures the shaking of the Earth caused by earthquakes, originally through the movement of a needle. When first introduced, the smallest detectable movement of the needle was 1 millimetre and the seismologists using these devices took this as a baseline minimum measurement.

Earthquakes vary greatly in size: the Earth is constantly suffering tiny tremors that are imperceptible and would never make the news. In fact, the earthquakes that measure close to the baseline level tend to occur around 100,000 times a year. On the other hand, 'one-in-every-ten-years' earthquakes have a level of shaking that is typically a million times as large. Differences in scale of that kind are hard to deal with in science and in news reporting. So, in 1935, Charles Richter devised a scale that could make comparison and analysis easier.

Richter classed the smallest detectable (at that time) earthquake, the base-line, as a magnitude 3 earthquake (he started at 3, sensibly anticipating that improvements in technology would in the future allow smaller earthquakes to be detected). From that starting point, every increase of 1 unit on the Richter scale means a 10-fold increase in the shaking amplitude.

This means that a magnitude 4 earthquake is 10 times the size of a magnitude 3 one, and a magnitude 5 earthquake is 10 times bigger again (making it 100 times bigger than the one of magnitude 3), and so on, all the way up to the million-times-greater earthquake, which is assigned a magnitude of 9. Interestingly, the frequency of occurrence of earthquakes tends to vary with the inverse of their size. So the million-times-greater earthquake happens one-millionth as frequently.[7]

This definition makes the Richter scale a log scale: the Richter magnitude can be directly calculated using the mathematical log of the seismograph's amplitude of shaking. So, once again, the use of a log scale transforms numbers that would be very hard to compare, because they differ so widely in magnitude, into a scale that is useful and manageable.

Every increase of 1 means a ten-fold increase in shaking, but in fact any set of equal-sized steps (they don't have to be 1 unit in size) in the Richter scale corresponds to the same proportionate increase in the shaking amplitude. If the step size is ½ (e.g. Richter 5 to Richter 5.5), the constant factor is approximately 3.16.[8] Taking two of these half-magnitude steps means multiplying by 3.16 twice, and that gives 10.

The Richter scale allows comparison of earthquakes from the very small to the very large on a consistent basis. It also means that we must be careful not to underestimate the effect of apparently small differences in the Richter scale.

For example, on 25 January 2016, between Spain and Morocco, there was a magnitude 6.3 earthquake. Then, on 6 February of the same year, in Taiwan, there was a magnitude 6.4 earthquake. Were these earthquakes of roughly the same size? (There was property damage in both cases but deaths only in the latter.) In fact, the 0.1 magnitude difference means a 26%[9] difference in shaking amplitude (and a 41% difference in energy release). So yes, 0.1 point difference in Richter does mean a significant difference in impact.

Landmark number A magnitude 8 earthquake is (on average) a once-a-year occurrence.

---

[7] The Richter scale relates to the amplitude of the shaking. The energy release in an earthquake increases even faster than the shaking amplitude. So a 1 point difference in Richter scale actually corresponds to a 31.6 times increase in energy release (31.6 is the square root of 1000).

[8] This number is the square root of 10.

[9] $10^{1/10} = 1.2589\ldots$

Turn it down!

The loudness of the ambient noise in a broadcast studio interior (when not in use) is reckoned to be 10 decibels (dB), a conversation at 1 metre distance is 50 dB, traffic at 10 m away can be as high as 90 dB, and a jet engine at 100 m is 130 dB. Damage to hearing can occur at prolonged exposure to sound at 85 dB, and instantaneous damage to your ears can be caused at 120 dB.

Decibels, which we use informally to describe volume of sound, actually measure the sound pressure level, or SPL. Although it's seldom used, the base unit for sound pressure level is actually the **bel**, named for Alexander Graham Bell, and a decibel is one tenth of a bel. So, the silent broadcast studio is 1 bel, the conversation is 5 bels, the traffic is 9 bels and the jet engine is 13 bels.[10]

Expressed in this way, it's easy to understand that the scale is very similar to the Richter scale: each increase of 1 bel means a 10 times increase in the SPL. So, the conversation that is 4 bels louder than the studio interior in fact has 10,000 times the SPL. Increasing by one decibel is like a 0.1 point increase in the Richter scale: it means a 26% increase.

By the way, just because a 10 dB difference means a sound is 10 times louder, don't think that means 20 dB is 20 times louder. Remember, that equal steps on a log scale translate to equal proportionate increases. So 20 dB louder means applying a 10 times increase twice, making for a 100 times increase.[11]

Landmark numbers
- Exposure to 100 dB for more than 15 minutes is unsafe (damaging to hearing).
- 100 dB is typical of power tools, lawnmowers, rock concerts, and football matches.

Ebony and ivory

A standard piano keyboard has 88 keys: 52 white and 36 black. In musical terms, each key (black or white) differs from the one to its left by playing a note

---

[10] What we perceive as sound volume is not actually the same as the value that scientific instruments measure. In fact, very roughly, our **perception** of volume **doubles** with every bel, every 10 decibels, while the sound pressure level multiplies by 10.

[11] And a 5 dB difference (in sound terms) is not half as much as a 10 dB difference, it's actually just 3.16 times louder (spot the square root of 10).

whose pitch is one semitone higher, and the pattern of white and black keys repeats every 12 semitones.

If you start with Middle C, right in the middle of the keyboard, and count 12 semitones to your right, counting both black and white keys, you'll have reached the next C note, called Treble C. This is one 'octave'[12] above Middle C. Take 12 more semitone steps, and you're at Top C, and that's two octaves above Middle C.

Heading the other way, if you start counting 12 semitones to the left from Middle C, you'll hit a C again (this is Bass C, one octave below Middle C). Go left for 12 keys more and you'll have reached Low C, two octaves below Middle C.

Now, a mathematician, looking at the musical description of a piano keyboard, might see it as a linear scale, counting semitones, where 12 semitones count up to an octave, and where the keyboard as a whole accommodates around 7½ octaves. The reasons behind the particular pattern of black and white notes are fascinating (and books have been written on just that!), but that is not relevant to us right now. Setting the pattern of black and white aside, you could almost think of the keyboard as a kind of measuring stick: just as there are 12 inches to the foot, so there are 12 semitones to the octave, but what this stick is measuring is not distance, but musical pitch on a chromatic[13] scale.

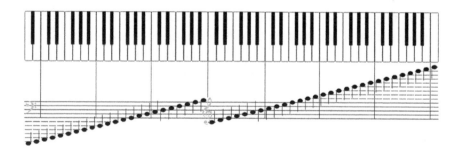

Now let's think about the physics of the thing.

When you strike Middle C, a properly tuned piano will sound a note with a frequency of around 261.6 hertz (Hz, or vibrations per second). That is, the strings in the piano will vibrate from side to side and back again, approximately

---

[12] You'd think an octave would mean 8 steps, but no. It's seven notes above the starting point (some notes involve whole tone steps, some involve semitones). But it's 'eight' if you count 'one' as your starting point. Musicians...

[13] 'Chromatic' in this context means valuing all semitones equally.

262 times every second. These vibrations (also at a frequency of 261.6 Hz) travel through the air and strike our ears where we hear Middle C.

Now strike the next higher C, Treble C. This time, the sound waves have a frequency that is twice as great—523.2 Hz. Top C? Twice as great again at 1046.5 Hz. For every octave up, the frequency doubles. Going down the piano, Bass C is 130.8 Hz, and Low C is 65.4 Hz, half and half again of the frequency of Middle C.

In fact, a healthy human hearing range is from 20 Hz to 20,000 Hz, a whopping thousand-fold increase in frequency. Just as with transistors and with earthquakes, when you're dealing with numbers that vary so much in scale, the best thing to do is to use logs. And that is exactly what a piano keyboard does: it converts equal proportionate differences in pitch (semitones and octaves) into evenly spaced distances on the keyboard. So a piano is in fact a device for producing reference frequencies that has an input mechanism based on a log scale. (Oh, you can play music on it too? A useful side-effect.)

Moving an octave up in pitch means a doubling of frequency (in terms of physics), and moving an octave down means a halving of frequency. What about the semitone? Musically speaking, 12 of them combine to make an octave. In terms of the physics, we want the effect of putting 12 semitones together to make a doubling of frequency. Remembering that every step on a log scale means a multiplication by a constant factor in the underlying system, the frequency change of one semitone must be whatever number can be multiplied by itself 12 times to give 2 (a doubling). That multiplier is $\sqrt[12]{2}$, the twelfth root of 2, which works out to around 1.06. As we go up the piano keyboard, each semitone step increases the frequency of the note by just under 6%. Do it 12 times, and the frequency has doubled.[14]

A piano, then, is just another log scale, where the huge difference in frequency between the lowest note (an A at 27.5 Hz) and the highest note (a C at 4,186 Hz) has been tamed into a manageable 87 equal semitone steps, grouped into seven-and-a-bit octaves.

> **Landmark number** The note that musicians generally tune to is the A above Middle C at 440 Hz.

---

[14] This 'equal' division of the octave into 12 semitones is correct for pianos that are tuned (as most are) using the 'equal temperament' system. Other instruments may use a 'just intonation' tuning where, for sound reasons, the 12 semitones in the octave do not each represent the same proportionate frequency ratio. Nonetheless, all 12 together still combine to make a doubling.

Log books and slide rules

Those who studied science and mathematics in the 1960s were probably in the last generation of students to use a 'log book'. This was a book of tables, providing a way of looking up the logarithm (to base 10) of numbers between 1 and 10.[15] Recall that on the log scale, two distances combined together will have a multiplicative effect. With a log book, to multiply two numbers together, you simply needed to look up the logs of each, add them together, then look up the 'antilog' (the inverse function). Division was done by subtraction of logs, and raising to a power by multiplying the log by the required power, before taking the antilog:

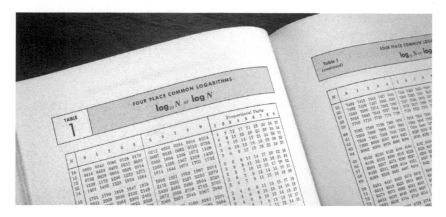

It's easy to forget, in these days when you can buy a scientific calculator for £5, just how useful and time-saving this was. The log book was a constant companion for anyone who needed to do calculations.

Even more iconic as a symbol of the engineer was the slide rule, which once again is based on the log scale. The basis of the slide rule is precisely the same as the number line we constructed at the start of this chapter for plotting large and small numbers together. A range of numbers running from 1 to 10 is marked along the length of the rule, in such a way that equal distances on the slide rule represent equal ratios between the marked numbers, which makes it a log scale.

The sliding mechanism allows two separate (but identically marked) scales to slide against each other, which makes it easy to add together two physical distances, and this corresponds to multiplying two numbers together. A skilled

---

[15] For values bigger than 10 or less than 1, we can scale: add 1 to the log to multiply by 10, and subtract 1 to divide by 10.

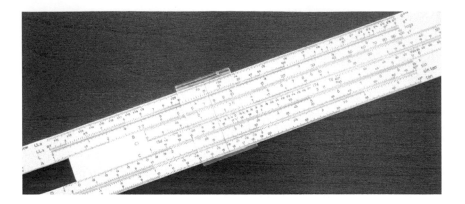

user of the slide rule could work very quickly to chain together a series of multiplications. Most slide rules also have a second scale running from 1 to 100, making operations such as squaring and taking of square roots easy to incorporate in the calculations.[16] Further scales added trig functions and exponentials. A good slide rule was a treasured possession.

So fundamental was the slide rule to the image of the scientist and engineer that science fiction writer Robert E. Heinlein took it for granted that in the future world he imagined, scientists and engineers would still be using them:

Dad says that anyone who can't use a slide rule is a cultural illiterate and should not be allowed to vote. Mine is a beauty—a K&E 20-inch Log–log Duplex Decitrig.

Robert A. Heinlein *Have Space Suit—Will Travel* (1958)

But the slide rule market utterly collapsed in the early 1970s, driven out by the introduction of pocket calculators that had far greater precision and were far easier to use. Something subtle may have been lost in this transition. A calculator is inherently digital, while a slide rule is an analogue device. Understanding a number displayed on a calculator is entirely an intellectual exercise, a matter of decoding symbols. However, with a slide rule, there is a perceptual, and a tactile quality. Its physical nature is a delight and the way it displays the answer to your calculation **in context** with a lot of other numbers played a role in developing an intuition about proportionality and ratio that the calculator can't replicate.

---

[16] The transparent part that slides up and down the slide rule with a fine hairline is used for matching up scales on the rule that are not adjacent, and is known as the cursor: this word is the origin of the cursor that I see on the computer screen before me, running ahead of my typing location, as I type these words.

I encourage you to find someone who owns a slide rule or get one for yourself—they're readily available on eBay—and play around with it. It drives home the deep message of log scales: that they reduce multiplicative operations to additive ones.

> **Landmark number** The halfway point on a slide rule's main scale is the square root of 10: approximately 3.16, and just a little bigger than π (pi). You can think of this as half an order of magnitude.

Mortality

Death will come to us all, but for most of us, we hope, not until we've lived a long, happy, and productive life. Actuaries and demographers study the chance of dying at various ages using what's called the 'mortality rate'. This is the chance that a person of a certain age will die in the next year of their life.

If you're aged	Chance of dying in the next year	
	Male	Female
25	0.055%	0.025%
50	0.31%	0.21%
75	3.34%	2.23%
100	36.2%	32.1%

From the table, you can see that every 25 years of life makes your chances of dying within one year (roughly) 10 times greater. So, once again we are dealing with data that varies greatly in scale, from less than 1 in a thousand to more than 1 in 10. This seems like good territory for the log-lenses.

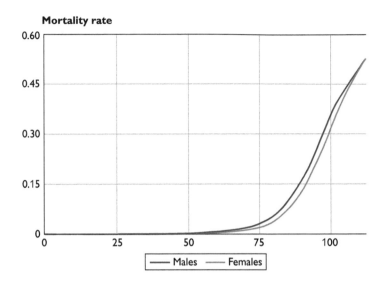

Look at the unadjusted graph above. The graph clearly shows some important things: that mortality below age 50 is very low in comparison with what comes later, that male mortality is greater than female after age 50, and that the rates come closer together in extreme old age. But this graph also hides information: it tells us nothing about mortality below age 50; nor does it give us any clue about the way in which mortality increases through the years.

Now let's look at that using the log of the mortality rate.

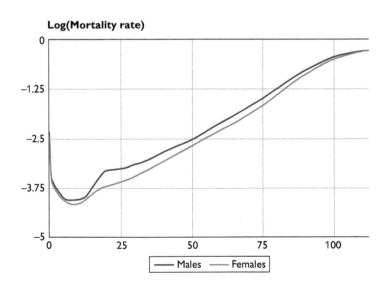

Bingo! We've suddenly got a lot more out of the data. For a start, we can see that from age 25 to 100 the line for both men and women is remarkably straight. As we saw with Moore's law, this means a more or less constant growth rate.[17] In fact, for every year that passes, your chance of dying in the next year increases by approximately 10%.

Another striking feature that the log scale brings out is the infant mortality 'hook' at the start of the curve. Even in modern Western society, the early years are risky ones.

Lastly, look at that acceleration and then levelling off of risk for males from around 15 to 25. This is known as the 'accident hump' and reflects young men's appetite for (or at least exposure to) risky behaviour in these years.

A much, much briefer history of time

These new-found log lenses are remarkably useful in all sorts of different contexts. As a final example, we can even use them to gain a little perspective on the immense stretches of times past, through history and prehistory. Let's place a few key events on a log-based time line that starts at a mere 10 years ago, but where each mark on the scale takes us 10 times further into the past.

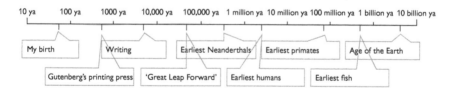

As we head into the next part of the book, we'll find more and more opportunities to use log scales to help us understand the huge numbers in the world of science.

---

[17] And that in turn suggests that the factors that cause death are ones that accumulate in a multiplicative way, steadily through life.

PART 3

The Numbers of Science

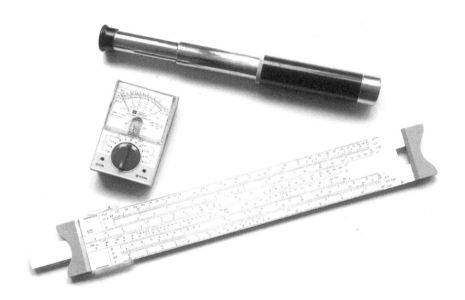

Thinking Big

If we succeed in giving the love of learning, the learning itself is sure to follow.

John Lubbock

Numbers for the sake of knowledge

So far, we've looked mostly at numbers that arise naturally from everyday human experience and activities. How much beer is in that barrel? How long is a ten-pin bowling lane? How heavy is a washing machine? How many people are there in this world? How far is it to Seattle? These numbers all help us make sense of the day-to-day world.

But life is more than beer and skittles. Our brains have the capacity and the desire to think beyond our daily lives—life would be dull if that were not so. We are inquisitive creatures. We would never have gained mastery of fire if our urge to know had not been stronger than our aversion to burnt fingers. So we asked the question 'How far is it to the Moon?' long before it there was ever a practical need for it, long before we ever contemplated sending a spacecraft there. It arose from sheer curiosity.

This is the pattern with science. The interesting questions are asked out of curiosity, for their own sake, from a desire for understanding. We want the knowledge that will come from the seeking and finding of answers. And sometimes—not always, but astonishingly often—the answers turn out to be useful. But it is only after the discovery has been made that practicality becomes part of the equation. Then the engineers can take on the task of realising the potential, turning science into solutions. When lasers were invented, they were a laboratory curiosity, a technology in search of an application. Now lasers are ubiquitous, but their many uses were never foreseen, and were never the motivation behind their invention.

This part of the book is about the big numbers that we find in science and that derive from the human urge to seek for knowledge and understanding of the world. Some of these numbers are hard to count or to measure directly. Many are not direct measurements but are estimates, or the result of calculations, which makes them harder to understand, or indeed to trust. Many are at scales so far removed from everyday life that they seem impossibly remote.

But big numbers in science are important. Not only are they often fascinating in themselves, but they help us to understand the world, and the universe, and our place in the universe, for good, or for bad.

Douglas Adams described an imagined device, the Total Perspective Vortex, in these terms:

Originally created by its inventor Trin Tragula as a way to get back at his wife (who was always telling him to get a 'sense of proportion'), the Vortex is now used as a torture and (in effect) killing device on the planet Frogstar B. The prospective victim of the TPV is placed within a small chamber wherein is displayed a model of the entire universe— together with a microscopic dot bearing the legend 'you are here'. The sense of perspective thereby conveyed destroys the victim's mind; it was stated that the TPV is the only known means of crushing a man's soul.

Now, Douglas Adams got so many things right, but in this case I think he was wrong. I find a sense of the scale of the universe, far from crushing the soul, is uplifting. I incline more to the attitude of Anatole France: 'The wonder is, not that the field of the stars is so vast, but that man has measured it.'

In the summer of 2015, the *New Horizons* space probe flew past Pluto and its moons. To me, once a geeky kid in the era of the *Apollo* moon landings, this seemed to be a human achievement of similarly epic proportions, and deeply inspiring in many ways. I mean, Pluto! Up close!

Every day, we tread our dusty paths. We seldom look more than a few footsteps ahead. But science lifts our eyes from the well-trodden trail, and not just to the horizon, but beyond, to the stars. Science inspires us by giving a vision of what could be, if only we were able to make it so.

Heavens Above

Measuring the Universe

The less one knows about the universe, the easier it is to explain.   **Leon Brunschvicg**

Which of these is the biggest?
- ☐ An Astronomical Unit (AU)
- ☐ Distance from the Sun to Neptune
- ☐ Length (circumference) of Earth's orbit around the Sun
- ☐ Halley's comet's furthest distance from the Sun (aphelion)

Reaching for the stars

[In] Astronomy, if you're getting within 10 orders of magnitude, you're okay.

Sir Michael Atiyah

How high the Moon?

We'll start this gently: the peak of Mount Everest is 8.85 km above sea level (5½ miles). What strikes me about that number is that it's not really very big. Not in any way to diminish the accomplishment of those who have climbed it, but if that height was turned on its side, into a horizontal distance, it's less than a 10 minutes' drive. Comparing it with the heights that humans have been able to create for themselves, it's no more than 11 times the height of Dubai's Burj Khalifa, the tallest skyscraper in the world.

At a slightly greater height, a commercial jet typically flies at around 13 km above the ground. Now, bear in mind that the radius of the Earth from centre to surface is approximately 6400 km. So you can visualise a passenger jet in flight as only barely skimming the surface, taking you a mere two parts in a thousand further away from the Earth's centre. These things are not very high at all.

Gliding in orbit way above the reach of any aircraft, the International Space Station (ISS) orbits the Earth at an altitude of 400 km. That's high, but still only around 6% further from the centre of the Earth than the ground is. That's known as a 'low Earth' orbit.

But not all satellites are in orbits as low as that. The higher the orbit, the further from Earth's gravity, the slower can be the satellite's rotation around the planet. In fact, those satellites that deliver television channels need to be in

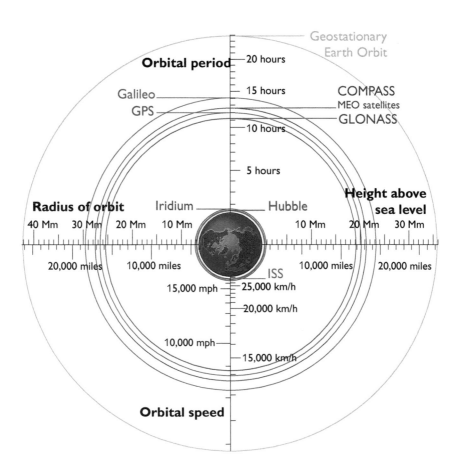

geostationary orbits. That is, they need to rotate around the Earth's equator at exactly the rate at which the Earth itself is rotating, which is to say one rotation per day. By contrast, the ISS whizzes around the Earth in 92 minutes. A geostationary satellite always appears to be in the same place when viewed from Earth (so that you can aim a static satellite dish at one of them). To remain in this kind of orbit, the satellite needs to be rather high up—35,800 km in fact, almost 90 times the altitude of the ISS. That's a distance that is starting to make the Earth itself look small.

So how far away is the Moon? Well, it varies, since its orbit is not perfectly circular, but, at its closest, the Moon is around 356,000 km from the Earth's centre, and, at its furthest, around 406,000 km. The average distance, which works out to 384,402 km, is sometimes used by astronomers as a unit in its own right, called the **lunar distance**, and this is more or less 10 times the height of the geostationary satellites.

Landmark numbers Here's a memorable sequence—very roughly:
- The International Space Station orbits at 400 km.
- A geostationary orbit is around 40,000 km.
- The equator is 40,000 km long.
- The Moon is more or less 400,000 km away.

And how big is the Moon compared with Earth? Its radius is 1740 km, which is around 27% that of Earth. That means its volume is around $1/50$ of Earth's volume. It also means that the diameter of the Moon, 3480 km, is less than the width (East to West) of Australia, which is approximately 4000 km.

Neighbourhood watch

That's just a walk around our block, relatively speaking: the Earth and its satellites. It's time to explore our wider neighbourhood: to do as Copernicus did and shift from an Earth-centred view of things to a Sun-centred one.

How far is the Sun from the Earth? Just about 150 million km, that's 390 times as far away as the Moon. How big is the Sun? Its radius is around 695,000 km, just under 400 times the size of the Moon. It's pure coincidence that the Sun is more or less 400 times further away than the Moon, and happens also to be about 400 times bigger, but this is a coincidence to relish. The only reason that we see the glories of a total eclipse of the Sun, with that spectacular corona visible, is because the apparent sizes of the Moon and Sun are so very nearly equal.

> **Landmark number** The Sun is 400 times further away, and 400 times bigger in diameter, than the Moon.

The Earth orbits the Sun at a distance of 150 million km. The speed of light being very close to 300 million m/s, this very memorably gives the time for light to travel to the Earth as 500 seconds, or 8⅓ minutes.

> **Landmark number** The Earth is 150 million km from the Sun, and that's equivalent to 500 light-seconds, or 8⅓ light-minutes.

The Earth travels around the Sun, making one revolution every year, a round trip of 940 million km (we can call it a bit less than a billion km—such a nice round number). So the Earth's speed around the Sun is just about 1 billion km per year, which works out to about 107,000 km/h.

> **Landmark number** The Earth travels nearly a billion kilometres a year on its trip around the Sun.

Two planets (Mercury and Venus) orbit the Sun more closely than Earth, and five (Mars, Jupiter, Saturn, Uranus, and Neptune) are further away. And that comparison serves to introduce a standard unit for talking about distances within the Solar System. What could be a more natural yardstick than the distance from Earth to Sun? So this distance (roughly 150,000 km) is referred to as an 'Astronomical Unit', or AU.

Here's a table of distances from the Sun showing both km and AU. I've also included the dwarf planets. The table also shows how the length of the planet's year (the orbital period) increases with distance from the Sun.

Body	Distance from Sun (km)	Distance from Sun (AU)	Orbital period (years)
Mercury	58 million	0.39	0.24
Venus	108 million	0.72	0.62
Earth	150 million	1.00	1
Mars	228 million	1.52	1.88
Ceres	414 million	2.77	4.6
Jupiter	778 million	5.20	11.9
Saturn	1.429 billion	9.55	29.4
Uranus	2.875 billion	19.22	83.8
Neptune	4.504 billion	30.11	164
Pluto	5.915 billion	39.53	248
Haumea	6.465 billion	43.22	283
Makemake	6.868 billion	45.91	310
Eris	10.166 billion	67.95	557

In May 2016, NASA's *New Horizons* space probe flew close to Pluto, reaching what you might have thought of as being the outer reaches of the Solar System. You'd be wrong, the Solar System is much larger than that. *New Horizons* is already heading for its next target, 2014 MU69, another object in the Kuiper Belt (a ring of objects between 30 and 50 AU from the Sun). *New Horizons* should reach it in 2019, by which time it will be about 6½ billion km from the Sun (43 AU).

The dwarf planet Eris (listed in the table) is one of a group of objects in solar orbit called the Scattered Disc, which occupies a band between 30 and 100 AUs distant from the Sun.

Landmark numbers
- Earth's orbit—1 AU from the Sun
- Saturn—Approximately 10 AU from the Sun
- Extent of the Scattered Disc—100 AU from the Sun

Where does the Solar System end? In a sense it's an arbitrary boundary, but there are two main indicators that astronomers use: the first is the region where the solar wind of particles blown out of the Sun merges with the 'sea' of particles of deep space. This limit is called the Heliopause, the boundary of the Heliosphere, at around 120 AU from the Sun, almost three times as far from the Sun as Pluto is.

Astronomers are still learning about this region, as data continues to be received from *Voyager 1*, launched in 1977. Nearly 40 years into its journey, *Voyager 1*, the human-created object that has travelled further from home than any other, has already moved through a number of transitional zones marking the edge of the Heliosphere, and is now almost 140 AU away from Earth. NASA have declared that, as of August 2012, *Voyager 1* is in interstellar space. Elvis has left the building.

This diagram shows the planets and the other features of the outer Solar System. Please notice that this is plotted on a **log scale**. Each number on the scale is 10 times further from the Sun than the previous number.

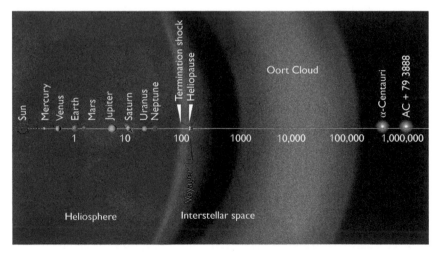

Reproduced with permission from NASA/JPL-Caltech

The most remote of the objects that orbit our Sun are part of the Oort Cloud, where comets go to rest between visits to the Sun (actually they're just slowly following their paths along very extended elliptical orbits). The Oort Cloud extends to at least about 50,000 AU away from the Sun, which is only a little less than one light-year—more of this in the next section.

The second candidate definition for the outer boundary of the Solar System is the distance at which the Sun's gravity is no longer the dominant gravitational force, but where the gravity of other stars starts to compete. Of course, this would not be a sharp boundary line, but at about 2 light-years away from the Sun, we reach a point where its gravity no longer dominates. At that point, we've certainly left behind all aspects of the Sun's influence. We're in outer space.

So it would be fair to say that the radius of the Solar System, although a very vague boundary, is at most about 2 light-years, and so we can think about the Solar System's zone of influence as a sphere at most 4 light-years across.

> **Landmark number** The Solar System is (estimating generously) 4 light-years across. Relatively speaking, though, the planets we know occupy only a very, very small central portion of that.

Light-years

It's confusing at first: you might expect a light-year to be a unit of time, but it really measures distance. But, from ancient times, one of the ways we've described distances is by saying how long that distance would take to travel at a certain pace: so we might describe a friend's home as being a 15-minute walk away, or a relative as living a two-hour drive away. We're making use of a common understanding of speed (walking speed, driving speed) to convert a time measure into a distance measure. Since the speed of light in space is a constant, and more than that, since it's the fastest speed that anything can travel, it's both natural and precise to measure distances in space by how far light can travel in a given time.

So to say that the Earth is 8⅓ light-minutes away from the Sun is no different in principle than saying that the station is 15 cycling-minutes away from my home. And light-seconds, light-hours, light-days are all perfectly good measures, but of course it's the light-year that is the important unit here: the distance light travels in a year.[1]

---

[1] Astronomers also like to use the **parsec** to measure distance to stars. That's about 3.26 light-years. It's used because it corresponds naturally to one of the ways astronomers calculate distance to the stars in our galaxy. Just as when you move your head from side to side, nearby objects seem

1 light-year: is that a big number? As ever, it depends on the context. It's about 63,250 Astronomical Units (AU). That means 63,250 times further than the distance from Sun to Earth.

And now we're starting to reach the point where (if we use kilometres as our base units), we need to join the scientists in how we talk about big numbers, using standard scientific notation. The following are equivalent:

- 1 light-year (ly) is approximately 9.5 trillion km.
- 1 light-year (ly) is approximately 9.5×10^{15} m.

> **Landmark number** 1 light-year
> - About halfway to the very furthest limits of the Solar System
> - Approximately 10 trillion km
> - Approximately 10^{16} m

Interstellar

The three stars nearest to the Sun are all part of the same star system, Alpha Centauri, which is 4.37 light-years away and is the third brightest visible 'star' in the sky. Of these three, the closest is a red dwarf, Proxima Centauri. We now know that Proxima Centauri has a planet in the habitable zone (with temperatures that could allow water to remain liquid).[2]

Next, at around 6 light-years distant, is Barnard's Star, a red dwarf, and not visible to the naked eye. The brightest star in the sky, Sirius, is around 8.6 light-years away. Procyon, 11.5 light-years away, is the eighth brightest star in the sky. (Canopus, the second brightest star optically, is much further away at 310 light-years.)

> **Landmark number** The star closest to the Sun is 4.37 light-years away.

Our Sun is part of a spiral galaxy that we call the Milky Way, and our immediate galactic neighbourhood is termed the 'Local Bubble', and is around 300

to move against the distant background (an effect called parallax), so, as the Earth moves around its orbit, nearby stars seem to change their position against the background of deep space. One parsec (parallax second) is the distance to a star whose position in the sky appears to change by one second of arc ($\frac{1}{3600}$ of a degree), when the Earth shifts in a perpendicular direction by one astronomical unit. In fact, in practice, the measurement would be based on observations made at suitable times of the year 6 months apart (meaning that the Earth had shifted by 2 AU), and then halved.

[2] It's remarkable that our Sun's very nearest neighbour seems to have a planet that could hold liquid water—this suggests that planets may be a very common feature of stars.

light-years across, 75 times our generous definition of the width of our Solar System. The Local Bubble forms part of the Orion–Cygnus arm of the Milky Way, one of the arms of the spiral galaxy. The arm is 3500 light-years across and around 10,000 light-years long.

The whole of the Milky Way is around 120,000 light-years across (recall for comparison that the Sun's sphere of gravitational influence is just 4 light-years across), and we are around 27,000 light-years from the centre of the galaxy. So we're little closer than halfway in.

> **Landmark number**
> Our 'Milky Way' galaxy is 120,000 light-years across.
>
> That's 400 times the width of our Local Bubble, and about 30,000 times the diameter of the Solar System.

Our galaxy is surrounded by a cluster of smaller satellite galaxies, but the nearest galaxy of comparable size to ours is Andromeda, which is 2.56 million light-years away, and is 220,000 light-years across (so almost twice as big across as our Milky Way). Both are part of the 'Local Group' of galaxies, which also includes the galaxy Triangulum and 51 smaller galaxies, and is around 10 million light-years across.

> **Landmark number**
> Diameter of our Local Group of galaxies—10 million light-years.
>
> That's about 80 times the width of the Milky Way.

Orders of magnitude

One of the ways we've managed the big numbers we've seen so far is by using a **divide and conquer** approach. We've sometimes chosen to break them each into two parts: a significand, a number in our comfort zone of 1 to 1000, and a scaling multiple (like 'billion'). The choice of a sensible unit, such as the light-year, helps further in thinking about these big numbers. But out in space, when we start to get to distances that are greater than a million light-years, this strategy is breaking down. Added to that, there really aren't very many familiar objects to help us establish **landmark** numbers for context.

The scales that we've been using are increasing rapidly, the sizes of the objects and the distances we're talking about are less and less precise, and so the significand part of the number is becoming less relevant. When numbers get this large, we're happy to retain a grip on the scale. As fine comparisons become less feasible, so our focus starts to shift to simply getting the scaling to the right order of magnitude. Think back to **log scales**, where each power of ten, each order of magnitude, could be represented by just one notch on a log scale. Out in intergalactic space, we've reached the point where the most useful strategy is to think in terms of counting the orders of magnitude.

So we might say that our galaxy's width, at around 10^{21} metres, is between four and five orders of magnitude greater than the width of the Solar System, at 4×10^{16} m. And the local group of galaxies is around two orders of magnitude bigger (in diameter) than our galaxy. Dealing with big numbers at this scale is mostly about knowing about the orders of magnitude.

Superclusters

Our galaxy is part of a local group, but at a much larger scale, it is also part of a supercluster of galaxies, a very large structure indeed, called the Laniakea Supercluster, which contains 100,000 galaxies, and which measures 520 million light-years across, which makes its diameter 52 times (almost two orders of magnitude) larger than that of our Local Group. It's thought that there are around 10 million superclusters in the observable universe. Superclusters themselves form even larger structures: filaments, walls, and sheets with dimensions stretching into the billions of light-years.

> **Landmark number** Diameter of the Laniakea Supercluster—520 million light-years, about 52 times the width of our Local Group

News from deep space

We've reached the point in this journey through big distances where simply counting orders of magnitude gives us our most meaningful comparison. Even this is not straightforward. In the course of writing this book, the estimate of the number of galaxies in the observable universe was revised from 100 billion to 2 trillion, increasing by a factor of 20. This was not a case of someone simply changing their opinion, but the result of 20 years of study of data from the Hubble Space Telescope, which revealed unprecedented detail. Now the successor to Hubble, the James Webb Space Telescope, has been built and is due to be launched in 2020. There's little doubt that the observations it helps us to make will trigger revisions to these big numbers too.

In 2015, the Laser Interferometer Gravitational-Wave Observatory (LIGO) detected for the first time gravitational waves emitted by the inward spiral and merger of two black holes. The event occurred 1.4 ± 0.6 billion light-years from Earth, three Laniakeas away. That's a big number, but still not as big as we can get.

The size of everything (well, everything we can see)

There is a limit to how far we can see into space: beyond that point, light would not have had long enough, from the start of the universe, to have reached us. Scientists have detected radiation from a time very shortly after the Big Bang (380,000 years), and have calculated that the location in space from which that radiation would have emerged is now about 46.5 billion light-years away from us (not only has the light been travelling towards us, but space itself has been expanding and taking the location of the origin away from us). This makes the diameter of the observable universe around 93 billion light-years, around 180 times the diameter of our supercluster.

So, counting orders of magnitude, and perhaps **visualising** them as measured out on a **log scale**, we can create the following chain:

> **Landmark numbers**
> - The universe is bigger than our supercluster by around 2 orders of magnitude.
> - Our supercluster is bigger than our Local Group by around 2 orders of magnitude.
> - Our Local Group is bigger than our galaxy by around 2 orders of magnitude.
> - Our galaxy is bigger than our local bubble by around 2½ orders of magnitude.
> - Our local bubble is bigger than the Solar System by around 2 orders of magnitude.
> - Our Solar System is bigger than Earth's orbit by around 5 orders of magnitude.

Taken together, this means that the diameter of the observable universe exceeds the diameter of Earth's orbit by 15½ orders of magnitude, which means around 3 thousand trillion. And it exceeds the diameter of our galaxy by a factor of a million.[3] And when it comes to distances, there are no meaningful numbers bigger than that.

---

[3] Of course, this is only the linear dimension. In terms of volume, we would have to cube this number to find out how many of our galaxies could fit into the observable universe. That comes to 1 million million million, or 1 quintillion. Perhaps better expressed: our galaxy, large as it is, is just one-quintillionth of the observable universe.

> **Landmark number**
> The observable universe is:
> • 93 billion light-years across
> • 6 quadrillion AUs across
> • Or 8.8×10^{26} m across

How far is a billion kilometres?

Here's a number ladder for space-related distances. Blank spaces are where there's just nothing sensible to add!

500 km	Altitude of the orbit of the International Space Station (400 km)
1000 km	Typical altitude of Earth-monitoring polar orbit satellite (1000 km)
2000 km	Diameter of Earth's Moon (3480 km)
5000 km	Mean radius of the Earth (6370 km)
10,000 km	Diameter of Venus (12,010 km)
20,000 km	Altitude of a geosynchronous orbit of a satellite around the Earth (35,800 km)
50,000 km	Diameter of Neptune (49,200 km)
100,000 km	Diameter of Saturn (116,400 km)
200,000 km	Diameter of Saturn's rings (282,000 km)
500,000 km	Distance from the Earth to the Moon (384,000 km)
1 million km	Diameter of the Sun (1.391 million km)
2 million km	Diameter of Sirius, the next-brightest (after the Sun) star (2.38 million km)
5 million km	
10 million km	
20 million km	
50 million km	Distance from the Sun to Mercury (58 million km)
100 million km	Diameter of Canopus, the third brightest star (after the Sun and Sirius) (99 million km)

	Distance from the Sun to the Earth (149.6 million km)
200 million km	Distance from the Sun to Mars (228 million km)
500 million km	Distance from the Sun to Ceres (414 million km)
1 billion km	Distance from the Sun to Jupiter (778 million km)
	Length (circumference) of Earth's orbit around the Sun (940 million km)
2 billion km	Diameter of VY Canis Majoris, the biggest star we can see (1.98 billion km)
	Distance from the Sun to Neptune (4.5 billion km)
5 billion km	Furthest distance of Halley's comet from the Sun (aphelion) (5.25 billion km)
10 billion km	Distance to Eris (dwarf planet) (10.17 billion km)
20 billion km	Distance to the Heliopause (17.95 billion km)
50 billion km	
100 billion km	
200 billion km	
500 billion km	
1 trillion km	
2 trillion km	
5 trillion km	Outer part of Oort Cloud (7.5 trillion km)
10 trillion km	A light-year (9.46 trillion km)
20 trillion km	A parsec (31 trillion km)
50 trillion km	Distance to Proxima Centauri (nearest star to the Sun) (39.9 trillion km)
100 trillion km	Distance to Sirius (brightest star in the night sky) (81.5 trillion km)
200 trillion km	
500 trillion km	
1 quadrillion km	
2 quadrillion km	Distance to Canopus (second brightest star in the night sky) (2.94 quadrillion km)
5 quadrillion km	

10 quadrillion km

20 quadrillion km

50 quadrillion km

100 quadrillion km

200 quadrillion km

500 quadrillion km

1 quintillion km   Diameter of the Milky Way (1.135 quintillion km)

2 quintillion km   Diameter of Andromeda (closest large galaxy to the Milky Way) (2.08 quintillion km)

5 quintillion km

10 quintillion km

20 quintillion km   Distance to Andromeda (24.22 quintillion km)

50 quintillion km

100 quintillion km   Diameter of Local Group of galaxies (95 quintillion km)

200 quintillion km

500 quintillion km

1 sextillion km

2 sextillion km

5 sextillion km   Diameter of Laniakea Supercluster of galaxies (4.92 sextillion km)

10 sextillion km

20 sextillion km

50 sextillion km

100 sextillion km

200 sextillion km

500 sextillion km

1 septillion km   Diameter of the observable universe (880 sextillion km)

The heaviness of the heavens

I have explained the phenomena of the heavens and of our sea by the force of gravity, but I have not yet assigned a cause to gravity.   Isaac Newton

How do you weigh a planet?

When we need to weigh 500 g of flour for a recipe, we'll use a kitchen scale, which will measure the force with which the Earth's gravitational field pulls on the mass of the flour in the scale pan. Happily, the scale will have been calibrated to correctly translate that force into a measurement of the mass of the flour.

Isaac Newton established the basic relationship that determines the gravitational force between two objects (it depends on the mass of each object and the distance between them), but there was a missing element. He knew the nature of the formula, but he didn't have a value for what was later named the Gravitational Constant (G), which was needed to do the full calculations implied by the formula. It wasn't until 1798 that Henry Cavendish was able to perform experiments that measured the gravitational force between massive spheres in a laboratory, and so was able to complete the work needed to use the formula in practice.[4] Cavendish's approach calculated the density of the Earth relative to that of water, reaching an answer of 5.448 (5.515 is the figure we would use today).

Knowing the volume of the Earth to be 1.08×10^{21} m^3, we can calculate that this means a mass of just about 6×10^{24} kg.

> **Landmark number** The mass of the Earth is 6×10^{24} kg.

That **is** a big number. We've not even started talking about anything beyond our blue Earth, and yet we have here a number that is bigger by a factor of 6000 than the diameter of our galaxy measured in metres, and bigger than all but two of the numbers we discussed in the section on distances. So this really is a challengingly big number. Is it even possible to form some sort of understanding of that number? How we can deal with the bigger numbers that are coming along?

We could restate the mass of the Earth as 6 million billion billion kg, or 6 septillion kilograms, but really, that scarcely helps. Let's do a very simple **visualisation**

---

[4] In fact, Henry Cavendish didn't express the formula in the way we do today. So, while he didn't calculate a value for G, his formulation was entirely equivalent.

exercise, just to help us unpack that figure and satisfy ourselves that it feels more or less plausible, at least to confirm that it's of the right order of magnitude.

We saw earlier in the book that the volume of the Earth was around 1 trillion cubic kilometres.[5] Cavendish found out that on average every cubic metre of Earth weighs around 5500 kg. And we know there are a billion cubic metres in a cubic kilometre. So taking these together, we can multiply 1 trillion × 1 billion × 5500. The trillion gives us twelve orders of magnitude, the billion gives nine and the 5500 gives three and a bit. So the product of all three has 24+ orders of magnitude. Our rough-and-ready calculation takes us to 5.5×10^{24}, and that compares well to the figure quoted above. Big numbers to be sure, but numbers that fit in with what we know about the Earth's size, and the density of rock and iron.[6]

But it's clear that it's not going to be easy to carry on thinking about planets and stars in terms of kilograms, or even tons. Perhaps a change of unit is called for? And that's just what astronomers do. For smaller planets, using 'Earth Mass' as a new unit makes sense, for the larger planets, 'Jupiter Mass' is suitable,[7] and for stars, we can use 'Solar Mass'.

Masses of the planets and Sun

Here's a table of masses of some notable objects in our Solar System:

Body	Mass (kg)	× Solar Mass	× Earth Mass	× Jupiter Mass
Sun	1.99×10^{30}	1.00	333,000	1,050
Mercury	3.30×10^{23}	1.66×10^{-7}	0.0553	1.74×10^{-4}
Venus	4.87×10^{24}	2.45×10^{-6}	0.815	2.56×10^{-3}
Earth	5.97×10^{24}	3.00×10^{-6}	1.000	3.14×10^{-3}
Moon	7.34×10^{22}	3.69×10^{-8}	0.0123	3.86×10^{-5}
Mars	6.42×10^{23}	3.23×10^{-7}	0.108	338×10^{-6}
Ceres	9.39×10^{20}	4.72×10^{-10}	1.57×10^{-4}	4.94×10^{-7}
Asteroid Belt	3.00×10^{21}	1.51×10^{-9}	5.02×10^{-4}	1.58×10^{-6}
Jupiter	1.90×10^{27}	9.55×10^{-4}	318	1.000

[5] Quick check for plausibility of **that** number: 1 trillion cubic kilometres is the volume of a cube 10,000 km on each edge. That would give a cubic version of Earth a 'circumference' of 40,000 km, which matches the length of the equator.

[6] Granite's density is around 2750 kg/m³, and iron's is around 7850 kg/m³.

[7] Increasingly, exoplanets—planets of suns other than our own—are being discovered. So these planetary yardstick masses are becoming useful as comparisons.

Body	Mass (kg)	× Solar Mass	× Earth Mass	× Jupiter Mass
Saturn	5.69×10^{26}	2.86×10^{-4}	95.3	0.299
Uranus	8.68×10^{25}	4.36×10^{-5}	14.5	**0.0457**
Neptune	1.02×10^{26}	5.13×10^{-5}	17.1	**0.0537**
Pluto	1.47×10^{22}	7.39×10^{-9}	2.46×10^{-3}	7.74×10^{-6}
Haumea	4.00×10^{21}	2.01×10^{-9}	6.70×10^{-4}	2.11×10^{-6}
Makemake	4.40×10^{21}	2.21×10^{-9}	7.37×10^{-4}	2.32×10^{-6}
Eris	1.66×10^{22}	8.35×10^{-9}	2.78×10^{-3}	8.74×10^{-6}

Some notable numbers from this table:

Landmark numbers
- The Sun has 333,000 times the mass of the Earth.
- The Sun has just over 1000 times the mass of Jupiter.
- The Sun's mass is very close to 2×10^{30} kg.
- Mars's mass is not much more than $\frac{1}{10}$ the mass of Earth.
- The mass of Earth is greater than those of Mercury, Venus and Mars put together.
- Uranus and Neptune taken together have just about a tenth of the mass of Jupiter.
- Eris is about $\frac{1}{8}$ heavier than Pluto.

The dance of the Earth and the Moon

As an aside, notice that the Moon has a mass of 1.2% of the Earth's, and think about where the centre of mass, the **barycentre**, of the Earth–Moon system is. If you drew a line between the centre of the Earth and the centre of the Moon, the barycentre would lie at a point approximately 1.2% along that line, which would make it around 4700 km from the Earth's centre, or around two-thirds of the distance from the centre of the Earth to the surface. We like to think of the Moon as rotating around a stable Earth, but both bodies rotate around this barycentre like an Olympic hammer-thrower preparing to throw. This means that the Earth's progress around the Sun is far from being a steady motion: it has a month-long wobble. Of all the planets in our system, only the Earth has a Moon that is so heavy—relatively speaking.

Comets and asteroids

In September 2016, the space probe *Rosetta* ended its mission of exploring the comet named 67P/Churyumov–Gerasimenko. This comet, 4.1 km along its longest dimension, has a mass of 1.0×10^{13} kg, that is, almost twelve orders of magnitude smaller than the Earth (a millionth of a millionth). Halley's comet, whose repeat visits have been tracked for centuries, is somewhat larger: 15 km in length and 30 times as massive as 67P.

The asteroid Ceres, appearing in the table above as a dwarf planet, is by far the biggest of the hundreds of thousands of asteroids that make up the belt. Ceres contains about ⅓ of the total mass of the Asteroid Belt, which in aggregate has a mass of approximately 3×10^{21} kg, or 4% of the mass of the Moon.

To our galaxy and beyond

The total mass of all the stars in our galaxy, the Milky Way, is estimated to be around 6×10^{11} Solar Masses (600 billion), or 1.2×10^{42} kg, and the total mass of all the visible matter in the observable universe is estimated to be of the order of 10^{53} kg, a further eleven orders of magnitude greater. This estimate excludes dark matter, whose nature is not yet known, and which is thought to outweigh ordinary matter by a ratio of five to one.

Astronomical densities

Earth is the densest of all the planets in our Solar System, at about 5500 kg/m³. (For comparison, water is 1000 kg/m³, iron is 7870 kg/m³, and granite is 2700 kg/m³). Mercury and Venus are slightly less dense, while Mars is considerably less dense at 3900 kg/m³. The Moon is still less dense than that, at 3340 kg/m³.

As you might expect, the densities of the gas giant planets are all considerably lower, ranging from Neptune at 1600 kg/m³ to Saturn at 700 kg/m³. That means Saturn is, on average, less dense than water and theoretically would float in water! The Sun's density is around 1400 kg/m³, much less than that of Earth.

The densest stars known to exist are neutron stars, which form when large stars collapse to a core of neutrons, with a density comparable to that of an atomic nucleus. Thus a neutron star might have a mass twice that of the Sun, and yet be only around 10 km across. That gives a density around 4×10^{17} kg/m³, fourteen orders of magnitude greater than that of the Sun.

What about black holes? Surely they must be the densest possible entities in the universe? The truth is, we just don't know. A black hole lives within a sphere known as the event horizon, and we have no evidence of what lies within that.

Good Heavens! I never knew that...

Diameter of **Uranus** (50,700 km) is
  4 × the diameter of the **Earth** (12,760 km)

Distance from the **Earth to the Moon** (384,000 km) is
  1 thousand × the length of the **River Thames** (386 km)

Diameter of **Saturn's rings** (282,000 km) is
  2 × the diameter of **Jupiter** (139,800 km)

Diameter of **Venus** (12,010 km) is
  5 thousand × the length of the **Epsom Derby horse race** (2.4 km)

Diameter of **Mars** (6790 km) is
  1 thousand × the length of the Oxford–Cambridge boat race (6.8 km)

Earth's orbit around the Sun (940 million km) is
  250 thousand × the length of the **Mississippi River** (3730 km)

Diameter of the **Sun** (1.391 million km) is
  400 × the diameter of Earth's **Moon** (3,480 km)

Length of the *Saturn V Apollo* launch rocket (110.6 m) is
  50 × the height of **Chewbacca** in the *Star Wars* movies (2.21 m)

A Bundle of Energy

Measuring the Spark

We wind a simple ring of iron with coils;
we establish the connections to the generator,
and with wonder and delight we note the effects
of strange forces which we bring into play,
which allow us to transform, to transmit and direct energy at will.   **Nikola Tesla**

Which of these is the biggest?
- ☐ Energy released by metabolising 1 gram of fat
- ☐ Energy in 1 gram of a meteor hitting the Earth
- ☐ Energy released by burning 1 gram of petrol (gasoline)
- ☐ Energy released by exploding 1 gram of TNT

Energetic numbers

— A burning match releases about 1000 J (joules) of energy. *Is that a big number?*

— A Snickers bar contains about 1.36 million joules. *Is that a big number?*

— A barrel of oil, when burnt, will release about 6 billion joules. *Is that a big number?*

Confused and confusing

Energy is a difficult subject. $E = mc^2$, said Einstein, teaching us that mass and energy are just different forms of the same thing. Now, quantifying matter and measuring mass (in other words, weighing) has been part of our human story, in many forms, at least since trading began. But it was only in the nineteenth century, with the work of James Joule, that different forms of energy were recognised as essentially the same thing. Even today, although energy is arguably even more fundamental to the make-up of our universe than mass is, and certainly a controversial topic in public life, it remains a confusing concept, and this is reflected in the ways we measure it.

This late arrival of energy as a unified concept is surely the reason why the measurement of energy is, frankly, such a mess. If you're looking for a tidy picture of measurement based on a single consistent base unit, and ever-increasing multiples arranged in a harmonious series of units, you won't find it here. Skip this chapter.

Instead, what you will find is a hotchpotch of measures each arising from, and each specific to, the human activity to which it relates. You will see different ways of measuring energy depending on whether it refers to:

- Food energy
- Electrical energy
- Fuel energy
- Explosive energy

In theory, all of these are united through the common SI unit of energy, the joule (J). In practice, though, joules are not what people have chosen to use to measure energy in day-to-day life.

Early measurement of energy

Of course, without knowing it, people have been trading energy for millennia—oil for heating and lighting, coal and peat for heating, even human beings themselves, bought and sold for their muscle. The cord of firewood, the barrel of oil: these have been our proxy measures for energy.

So there is no ancient or historic human-scale measure of energy, except perhaps for a day's labour.[8] The first actual quantification of energy was not in fact a measurement—there would have been no suitable measuring device available—but a calculation, by Gottfried Leibniz in the late seventeenth century. Leibniz had noticed that in many mechanical systems, a particular calculated quantity (based on the mass and the square of the speed of the bodies) was conserved. He termed this quantity *vis viva*, the force of life, and, apart from a constant scaling difference, it is what we would today call kinetic energy.

---

[8] Roughly speaking, 6000 kJ.

Because energy comes in so many different forms (kinetic energy, potential energy, chemical energy), there is no standard and consistent means of measuring it. James Joule, though, working in the mid-nineteenth century, is regarded as the first to recognise and measure how potential energy is transformed into heat energy, which he detected as a rise in temperature. His device used a raised weight, which, when released, descended, driving a paddle mechanism to agitate water. This created kinetic energy, which was captured as heat in the water and caused its temperature to increase. In his honour, the unit of energy in the SI is called the joule (J). But how big is a joule?

How big is a joule?

This is where I risk losing you. The definition of a joule is not straightforward. It's not based on body parts or barrels of goods, or anything that we would be directly familiar with from everyday experience. It is a **derived unit**, which means that it is defined in terms of other more basic units.

A joule is the amount of energy you spend when you exert a force of 1 newton over a distance of 1 metre. But what's a newton? Who invited Newton to this party? Well, you can't have Joule without Newton. A newton is the SI unit of force, and is also a derived unit, and is defined as the force needed to accelerate one kilogram of mass at the rate of one metre per second per second.

This isn't helping, is it? I can imagine that your eyes glazed over when you read that last paragraph. I suspect that this definition isn't giving you any sort of intuitive understanding. Let's try a different tack.[9] Here's a list of things that have an energy of about 1 J:

- The energy expended when a 100 g tomato falls 1 m, and fails to bounce. Plop!
- The heat energy needed to raise the temperature of 1 ml of water by about ¼ of a degree Celsius.[10]

---

[9] For those who want to stick with this, imagine a force (such as your hand) pushing a 1 kg mass on a frictionless surface from a stationary start for a distance of one metre, keeping the force constant so that the mass accelerates smoothly. If you can do this in a time of about 1.4 seconds, then by the time you reach the end of that metre's distance, the mass will now be moving at a rate of 1.4 metres per second. You'll have imparted 1 joule of kinetic energy to the mass. Still not exactly intuitive, is it?

[10] This links the joule to an alternative definition of energy, the calorie, which will raise the temperature of 1 g of water by 1 degree, and echoes Joule's original experiment.

- The amount of electricity required to light a 1 watt LED for 1 second—now James Watt's turned up—and a 1 watt LED is good enough for a reading lamp.

So, as you can see, the joule is a rather small unit of energy. Human-scale? Yes, but barely, and rather too small to be very useful. An AA battery contains approximately 10,000 joules of energy: that's a biggish number for a smallish energy store.

So if the joule is too small to be very practical as a unit, let's see if there is something a little more suitably sized. The standard unit for recording mains electricity consumption is the kilowatt-hour (kWh), one of which is equivalent to 3.6 million joules. If you boiled your kitchen kettle (rated at 3000 W) four times in the day, for 5 minutes each time, you'd use 1 kilowatt-hour.

There. That feels like a sensible, human-scale measurement of energy, the kilowatt-hour, which in the UK in 2016 would cost around 15p. In the UK the energy regulator Ofgem publishes figures for electricity consumption of typical households. Their profile for a 'medium' household on a conventional plan uses 3100 kWh per year, which equates to around 8.5 kWh of electricity per day. Most UK households rely on mains gas for heating and hot water, and typically, in energy terms, their gas consumption is around four times the electricity usage, meaning that a such a household might typically use a bit more than 40 kWh of energy per day, taking gas and electricity together.[11]

Energy in foods

We also consume energy in a literal sense, to fuel our bodies with food. Conventionally, we measure the amount of food energy by counting calories. The scientific definition of a calorie is the energy needed to raise the temperature of 1 g of water by 1 °C. The exact amount of energy needed for this task depends on factors like air pressure and starting temperature, which means there are several different possible definitions of a calorie. The most commonly used is the thermochemical calorie, equivalent to 4.184 joules.

But the situation gets even more confusing, since the calorie is a rather small measure, and those involved with nutritional energy have adopted the convention

---

[11] Incidentally, the British Thermal Unit (BTU) is used to measure fuel energy content in North America, but ironically is no longer used in Britain. It is an imperial equivalent of the kilocalorie, being the amount of energy required to heat one pound of water by one °F.

of measuring food energy in **kilo**calories, but still calling them 'Calories' (capitalised). These are the familiar Calories that should be counted if your weight is to be watched.

Since a food Calorie is 1000 times more than a scientific calorie, it follows that it is the energy needed to raise the temperature of 1 kilogram of water by 1 degree. So if your kettle holds 2 litres of water (weighing roughly 2 kg), and you boil it to raise the temperature from 20 to 100 degrees, that will require 160 Calories, which is equivalent to around 670 kilojoules (kJ).

You won't have any trouble finding Calorie-counting lists of foods on the Internet, but here are a few items, just to establish a context:

- Chicken salad sandwich—250 Calories
- Glass of orange juice (200 ml)—90 Calories
- Snickers bar (64.5 g)—325 Calories

It's one thing to think about how much energy is contained in the food you eat, another to think about how much you expend.

The current recommended daily consumption of Calories is 2500 for a man, 2000 for a woman. We're used to seeing tables of how much energy can be burned off by particular energetic exercises, but the main way our bodies burn off energy is pure survival—simply to keep our basic body processes ticking over, stopping us from dying. These basic metabolic processes consume the bulk of the energy we expend on a daily basis (even without doing any exercise). The so-called Basal Metabolic Rate, or BMR, can be more than three-quarters of the daily intake of calories. This means the effect of, say, doubling the exercise you take is not as great as you might think: if 75% of your energy is accounted for by the Basal Metabolic Rate (and the remaining 25% by exercise), then doubling the exercise you take will lead to burning only 25% more Calories.

Energy in fuels

How much energy is there in a litre of petrol? The energy density of petrol is 34.2 megajoules per litre (MJ/L), which is approximately 10 kWh per litre, and incidentally is more or less equal to the usable energy contained in an equal volume of animal or vegetable fat. (That's enough energy to boil your kettle four times a day for 10 days, and about a quarter of the amount needed to run the average household for a day in the UK.)

Here are some measures of energy density (in kWh/kg) for petrol and other fuels:

Lead–acid battery	0.047 kWh/kg
Alkaline battery	0.139 kWh/kg
Lithium-ion battery	Up to 0.240 kWh/kg
Gunpowder	0.833 kWh/kg
TNT	1.278 kWh/kg
Wood	4.5 kWh/kg
Coal	Up to around 9.7 kWh/kg
Ethanol	7.3 kWh/kg
Edible fat	10.3 kWh/kg
Petrol	10.9 kWh/kg
Diesel	13.3 kWh/kg
Natural gas	15.4 kWh/kg
Compressed hydrogen	39.4 kWh/kg

What's your energy consumption?

The amount of energy we're each accountable for varies considerably from country to country. You shouldn't be surprised to learn that living in a cold climate would drive up energy usage. You might also expect that having a cheap source of energy close to hand would boost energy consumption, and it's these two factors that mean Iceland is among the countries with the highest per-capita energy consumption. In 2014, their energy consumption was the equivalent of burning 59 litres of petrol per person, per day. Of course, the Icelanders are not burning petrol: a great proportion of Iceland's energy usage is from renewable sources. Almost all its electricity comes from hydroelectricity and geothermal generation, and 90% of its heating needs are met directly from geothermal sources.

The energy usage of the USA was in 2014 equivalent to around 23 litres of petrol per person per day, and in the UK the figure is just over 9 litres. For China in that year it was around 7.4 litres, and, to take an example near the bottom of the table, in Pakistan it was 1.6 litres per person per day.

How much energy does the whole world use? In 2014, the total energy consumption throughout the world came to roughly 14,000 Mtoe. Uh-oh, 'Mtoe', yet another unit: this one is 'million tons of oil equivalent', which is how energy is measured once you get to really big numbers.[12] So 1 toe is 42 gigajoules, or 11.62 megawatt-hours (MWh).

Working this through, the world's energy consumption in 2014 was 14,000 × 1,000,000 × 11,620 kWh. That comes to 1.627×10^{14} kWh, for the whole world. Let's do a **cross-comparison** to check that for reasonability, and using an approach based on our **rates and ratios** technique.

Divide that total annual energy consumption by a world population of 7.24 billion to get a per-capita figure of 22,470 kWh per person for the whole of 2014, or 62 kWh per person per day, equivalent to £9.30 worth of energy at UK rates, or boiling your kettle 248 times. That seems rather a lot, but this figure is not just domestic energy. This includes all the other ways you use energy directly or indirectly, such as at work, in taking the train, or in going to a concert. It also includes all the energy used in the world on your behalf, in factories that make the goods you buy, and all manner of commercial activities. It also includes all the energy lost through power transmission, inefficiencies of generation, and other wasted energy, which has been estimated to be as high as 25%.

> **Landmark number** The average amount of energy consumed per person per day (in 2014) was around 60 kWh, roughly the energy equivalent of burning 10 litres of petrol.

But perhaps that average figure is a little misleading. It masks a very high degree of variability in the energy usage in different parts of the world, where the average Icelander consumes 36 times as much energy as the average Pakistani. Averages can be misleading, something we will pay much more attention to later on in the book.

Temperature

Temperature is not in itself energy, but an outward sign of inner heat energy. However, unlike energy, temperature is something that we directly sense in the world. Living organisms, their metabolisms, and most chemical reactions are very sensitive to temperature. For this reason, our bodies continually, if unconsciously, regulate our temperature, and when this mechanism falters, it's a sign that we

---

[12] Incidentally, there is also a 'ton of coal equivalent' unit (tce).

are very ill indeed. Precise measurement and control of temperature is essential for the manufacture of substances as different as steel and chocolate. The language tells us as much—both of these need to be 'tempered' by reaching the correct 'temperature'.

In the past, temperature was simply an impression of how hot or cold something was, a quality naturally sensed and easily understood, and requiring skill of judgement on the part of chefs and smiths. Temperature was not put on a scientific footing, or explained physically until the theory of statistical thermodynamics had been formulated: that the temperature of a substance is really a measurement of the average amount of vibration of the constituent molecules. Although we have a direct sense of temperature, it is very subjective, and to measure it objectively for scientific purposes relies on the physical fact that many substances expand and contract in a predictable way in response to a change in temperature.

The first proper measures of temperature were based on setting reference points and calibrating a scale between these reference points. Daniel Fahrenheit invented the mercury-in-glass thermometer in 1714, and published his scale for measuring temperature ten years later. He based the zero for his scale on the freezing point of a brine solution, designated the freezing point of water as 32 °F, and took as a third reference point the temperature of a healthy human as 96 °F (three times the freezing/melting point of water). Later, the freezing brine and the healthy human points were discarded from the scale, and it was recalibrated to incorporate the boiling point of water as 180 °F above the freezing point, namely 212 °F. All other intermediate temperatures on the scale rely on the assumption that the expansion of mercury in a thermometer is proportionate to the temperature change.

The Celsius (also known as centigrade) scale used the arguably more rational 0 °C for the freezing point of water and 100 °C for the boiling point. Both of these scales allow the temperature measurement to take negative values, something that is relatively uncommon in the world of everyday measurements.

Temperature is a statistical measure of the vibrational heat energy in a substance, and, knowing this, we can see the possibility of this vibrational energy dropping to zero. This leads to the logic of a temperature scale where the zero point is based, not on freezing water, but on this point of zero activity. Such a scale indeed exists, the Kelvin scale, and its zero is known as absolute zero. The size of each kelvin is the same as a degree Celsius—and this makes the freezing point of water equal to 273.15 K.

Once again, we see a tension between everyday measurements and scientific measurements. For scientific work, especially at very low temperatures, it makes no sense to use anything other than kelvins,[13] while for day-to-day life, to use the Kelvin scale would be absurd, and Celsius and Fahrenheit remain the scales of choice.

Rule of thumb

°C = 5 × (°F − 32)/9

°F = 32 + 9 × °C/5

K = °C + 273.15

Landmark numbers

- At −40 degrees, Celsius equals Fahrenheit
- 10 °C = 50 °F (a cool day)
- 25 °C = 77 °F (a pleasantly warm day)
- 40 °C = 104 °F (a serious fever, seek medical help)

Some other notable temperatures:

- Melting points and boiling points
  - Helium: 0.95 K and 4.22 K (−272.2 °C and −268.93 °C)
  - Nitrogen: 63.15 K and 77.36 K (−210.00 °C and −195.79 °C)
  - Carbon dioxide: (sublimes at)[14] 194.65 K (−78.50 °C)
  - Mercury: −39 °C and 357 °C
  - Tin: 232 °C and 2602 °C
  - Lead: 328 °C and 1749 °C
  - Silver: 962 °C and 2162 °C
  - Gold: 1064 °C and 2970 °C
  - Copper: 1085 °C and 2562 °C
  - Iron: 1538 °C[15] and 2861 °C
- Ignition points
  - Diesel: 210 °C
  - Petrol (gasoline): 247–280 °C
  - Ethanol (alcohol): 363 °C

---

[13] Note that the units of the Kelvin scale are (and have been since 1968) simply kelvins, not 'degrees Kelvin' as they had been.

[14] This is the 'dry ice' that turns directly from solid to vapour and produces the sheets of manufactured mist popular in rock (and other) concerts.

[15] The very high melting point of iron is one of the reasons that the first use of iron for toolmaking, marking the start of the Iron Age, came so long after the use of bronze, which melts at around 950 °C. Our ancestors had to get their fires more than 500 °C hotter.

- Butane: 405 °C
- Paper: 218–246 °C[16]
- Leather/parchment: 200–212 °C
- Magnesium: 473 °C
- Hydrogen: 536 °C
- Others
  - Domestic fire: around 600 °C
  - Blacksmith's smithy: between 650 °C and 1300 °C
  - Steel blast furnace: up to 2000 °C
  - Aluminium-based fireworks: 3000 °C

Big and small bangs

Comparing energy in different forms is fraught with difficulty. A Snickers bar contains energy amounting to 1.36 million joules. This is equivalent to the energy in 0.45 kg of gunpowder, but the effects of releasing that energy are very different, and that's all to do with the rate at which they burn.

Gunpowder has an energy density of around 3 megajoules per kilogram—so igniting 1 g of gunpowder would release 3000 J, or around a third of the energy in an AA battery. In fact gunpowder's energy density is less than $^1/_{10}$ of the energy density of petrol. What makes it effective as an explosive is the very high rate at which it releases that energy.

A rifle bullet, to take a specific example, a 5.56 NATO cartridge, has a muzzle energy of 1800 J. When it strikes, that 'tiny' amount of kinetic energy is very rapidly released into the target, with great destructive force operating for a very brief time.

Display-quality fireworks contain around 100 g of explosive 'flash powder', which has an energy density of about 9.2 MJ/kg, releasing around 1000 kJ of energy, around 500 times as much as the rifle bullet.

Trinitrotoluene (TNT) has a lower energy density than the flash powder, at around 4.6 MJ/kg, but TNT has become the standard against which explosions are measured. The amount of energy released when a ton of TNT is exploded is widely used as a yardstick for comparing explosions, and the standardised value is taken to be 4.184 GJ. For example, a Tomahawk Cruise missile delivers an explosion that is approximately equivalent to half a ton of TNT, while the

---

[16] 424–475 °F. This range does indeed include the temperature used by Ray Bradbury for the title of his dystopian novel *Fahrenheit 451*, the temperature at which paper catches fire, where the 'firemen' were the ones burning books.

atomic bomb dropped on Hiroshima was equivalent to 15 kilotons. Bombs developed at the height of the Cold War could deliver up to 50 megatons, equivalent to over 100 petajoules.[17]

But nature outdoes humankind when it comes to destructive energy. The volcanic explosion of Mount Krakatoa in 1883 is estimated to have had an energy of 200 megatons and affected the world's weather for years.[18] And every minute, the Sun beams down something like 5 exajoules (5 quintillion, 5×10^{18} joules) to the Earth's surface—that's 1200 megatons—6 Krakatoas every minute.

Energy number ladder

1 J	Impact of a 100 g tomato falling 1 m
300 J	Kinetic energy of a person jumping as high as they can
360 J	Kinetic energy of a world-class javelin throw
600 J	Kinetic energy of a world-class discus throw
800 J	Energy required to lift an 80 kg person 1 metre against Earth's gravity
1400 J	Unfiltered solar radiation received in 1 s by 1 m² of surface area at the Earth's orbit
2300 J	Energy to vaporise 1 g of water
3400 J	Kinetic energy of a world-class hammer throw
4200 J	Energy released by explosion of 1 g of TNT = energy in 1 food Calorie
7000 J	Muzzle energy of a 0.458 Winchester Magnum
9000 J	Energy in an AA battery
38,000 J	Energy released by metabolising 1 g of fat
45,000 J	Energy released by burning 1 g of petrol
300,000 J	Kinetic energy of a 1 ton vehicle (small car) travelling at 90 km/h
1.2 MJ	Food energy of a Snickers bar (280 Calories)
4.2 MJ	Energy released by 1 kg of exploding TNT
10 MJ	Recommended daily energy intake of a reasonably active person.

[17] A petajoule is 10^{15} J, a million billion joules.
[18] It was also called the loudest noise ever heard by human beings.

Smash! Bang! Wallop!

Energy released by exploding **1 g of TNT** (4.2 kJ) is
about as much as energy in **1 food Calorie (kcal)** (4.2 kJ)

Energy released by metabolising **1 g of carbohydrate** (17 kJ) is
5 × kinetic energy in an **Olympic hammer throw** (3.4 kJ)

Energy released by metabolising a **shot of espresso** (92 kJ) is
40 × energy to **vaporise 1 g of water** (2.3 kJ)

Energy released by **burning 1 g of petrol** (45 kJ) is
5 × energy in an **AA alkaline battery** (9 kJ)

Energy in an **Olympic shotput throw** (780 J) is
about as much as the energy needed to **lift an 80 kg person** 1 m (780 J)

Energy released by metabolising a **pint of beer** (764 kJ) is
2.5 × energy released by metabolising a **50 g egg** (308 kJ)

Energy released by metabolising a **glass of wine** (450 kJ) is
10 × energy released by burning **1 g of petrol** (45 kJ)

Energy released by burning **1 g of petrol** (45 kJ) is
25 × kinetic energy of an **M16 rifle bullet** (1.8 kJ)

Energy of the explosion of **Krakatoa** (837 TJ) is
4 × yield of the **largest nuclear weapon** ever tested (210 TJ)

Bits, Bytes, and Words

Measurement for the Information Age

A bit, the smallest unit of information, the fundamental particle of information theory, is a choice, yes or no, on or off. It's a choice that you can embody in electrical circuits, and it is thanks to that that we have all this ubiquitous computing.   **James Gleick**

Which of these had the most computer memory?
- ☐ First Apple Macintosh Computer
- ☐ First IBM Personal Computer
- ☐ BBC Micro Model B Computer
- ☐ First Commodore 64 Computer

Numbers about information

All of the books in the world contain no more information than is broadcast as video in a single large American city in a single year. Not all bits have equal value.

Carl Sagan Cosmos (1980)

— My computer uses a 64-bit architecture. *Is that a big number?*

— *Moby Dick* has 206 thousand words. *Is that a big number?*

— I just bought a 4-terabyte hard disk. *Is that a big number?*

Rosetta

The Rosetta Disk is a three-inch nickel disk covered in engraved script. Starting from the edge, eight texts, in eight languages, spiral inwards, getting ever-smaller as they go. The English text reads:

Languages of the World: This is an archive of over 1,500 human languages assembled in the year 02008 C.E. Magnify 1,000 times to find over 13,000 pages of language documentation.

Photograph taken by Kevin Kelly, provided courtesy of The Long Now Foundation's Rosetta Project

The Disk has been produced by the Rosetta Project, which aims to explore what it would take to create and preserve a library that would endure for 10,000 years. It is an exercise in long-term thinking. The time span contemplated is around twice as long as the oldest writing we have today, and the aim of the Rosetta Disk is to provide the key that future users of the library, or indeed archaeologists exploring the remains of our civilisation, will need to unlock any written scripts from what will to them be ancient times. It will fulfil for them the role that the Rosetta Stone fulfilled for us in unlocking the Egyptian hieroglyphs.

The inward-spiralling texts are meant to convey to the finder that magnification is the way to discover all that the disk contains. And, unlike a CD or a DVD, optical magnification is the only technology needed to disclose all the contents. When enlarged, the viewer will see images of conventional book pages—human-readable. Many copies of the disk will be produced, following a principle the project has dubbed LOCKSS—'Lots Of Copies Keeps Stuff Safe'.

It's a remarkable project, one that recognises the importance of information in our culture. It may never be needed. Humanity and our cultures may continue to thrive and decline without any catastrophic discontinuity. Perhaps there will never again be any more dark ages. But 10,000 years is a long time.

Information

We've discussed the measurement of space and time, of mass and energy. This section is about something that is rather less physical, less tangible, but no less important for all that. Indeed, for many of us, our daily occupations are dedicated to finding, manipulating, and propagating this ethereal thing we call information.

Information is about choices. Around 5000 years ago, a Sumerian scribe took his wedge-shaped reed and pushed it into a tablet of soft clay to make imprints (using a set of conventions that that we now call cuneiform writing) that would signify '50 bushels of grain'. Perhaps he was communicating with a buyer or a seller, or perhaps he was simply making a record of this year's harvest from a particular field, so that next year he could make a comparison. Whatever the reason, when he inscribed his marks, he was creating information. Now, matter and energy are subject to laws of conservation. They cannot be created from nothing, and they cannot be destroyed entirely. But information is not the same. It requires no more than a medium to carry the text, a way of marking, and a meaning to be expressed. While matter and energy are never entirely destroyed, information can be. It requires no more than a little jumbling-up.[19]

Information is reliant on symbol-making. Whether the symbols take the form of scratches on an antler, marks in clay, ink on vellum, print on paper, or transistors in on-and-off states in a computer chip, it all counts as information, and it can all, at least in principle, be encoded and stored using information technology.

In everyday use, the term 'information' is rather vague and imprecise, meaning some kind of dormant communication. It's hardly the sort of thing that you might think could be made precise, still less be measured. But at the very dawn of our information age, a brilliant telecommunications engineer called Claude Shannon, working for Bell Telephone Laboratories, developed a theory of information that now underpins much of our modern digital world. His theory

---

[19] Physicists would contend that, at a quantum level, information is subject to its own laws of conservation. At a human level though, when the tiles from a completed Scrabble™ game are tossed back into the bag, information is destroyed.

sought to provide precise definitions of the concepts associated with communication, and this allowed us to measure this intangible information. Today, when you choose how much memory capacity you need for your new computer, or complain about your broadband speed, you've Shannon to thank for even being able to express those concepts.

Measuring information

In 1948, Shannon published *A Mathematical Theory of Communication*, a paper that introduced the use of the word 'information' in a formal and rigorous sense. In the formalism of Shannon's theory, when a message is received, that message conveys information by resolving some measure of uncertainty at the receiving end. 'Uncertainty' was itself a concept that needed to be formalised. Flip a coin. Heads or tails, there's your atomic unit of uncertainty, an unresolved two-way choice, and Shannon saw that the role of communication was to resolve this uncertainty through the delivery of information.

The quantity of information carried by a message is a measure of how much uncertainty is removed when a message is received. The smallest unit of information is the **bit** (binary digit), which simply indicates which of two equally likely choices has been made.

The American revolutionary Paul Revere said he'd announce the arrival of British forces by placing lamps in the church tower: one lamp if the British were coming by land, two if by sea. This was a two-way choice, and his message would remove uncertainty in regard to that binary option. Shannon would have said that Revere's message would deliver just one **bit** of information.

Just as atoms can combine to form molecules, so these atomic bits can combine to form larger structures, longer messages carrying more information and resolving greater uncertainties. Suppose that, instead of 'by land' or 'by sea', Paul Revere needed to signal one of four possible choices, for example whether the British were coming from East, West, North, or South. This is now a four-way choice—so there are more possibilities that the message needs to discriminate between. You could signal a choice like this in many ways—one, two, three, or four lanterns, or through the use of coloured filters, but Shannon's approach tells us that any such multi-way choice can always be made equivalent to using a combination of two-way choices. In this case, we can create a decision tree composed of binary choices, where every path involves two decision points.

So, the first choice might split the four alternatives into two pairs: {North or South} versus {East or West}, and the second choice would further refine that

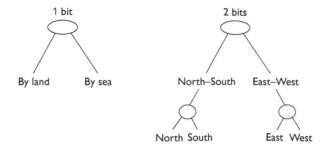

choice to a single direction. To find the answer means processing a message containing two elements: two bits.

If there were eight equally likely messages he wanted to choose from, Paul Revere would have needed a code capable of sending three bits of information: three two-way choices made one after another lead to eight possible different outcomes, in just the way that tossing a coin three times can lead to eight possible sequences of heads and tails. So a coding system that chooses from eight possibilities encodes three bits of information in every message.[20]

This is exactly what happens when one computer communicates with another. A sequence of bits is transmitted, and because common standards are employed, the meanings of these bitstreams are understood. They express choices between alternatives using agreed conventions, such as which character should be displayed, or what colour pixel to paint on the screen.[21] But before we discuss computers, let's look at a different example of an encoding scheme.

Semaphore is another early way of sending messages over long distances. Each of the semaphore flags can be in one of eight different positions, and so information theory tells us that one flag should be able to encode three bits of information. There are two flags: so, in theory, if the flags could be visually distinguished, six bits of information could be sent, for $2^6 = 64$ possibilities.

However, eight of the positions have both the flags in the same position, and only one of these eight positions is actually used in the code (both flags down,

[20] The number of possibilities can be calculated as 2^b, where b is the number of bits of information available. Turned around, this means that the number of bits of information is $\log_2 n$, where n is the number of equally likely possibilities.

[21] At the lowest level, all these information processes are extremely simple, absurdly so. The ability to represent complexity comes from the way in which huge numbers of these atoms of information and computation are assembled into ever-larger structures.

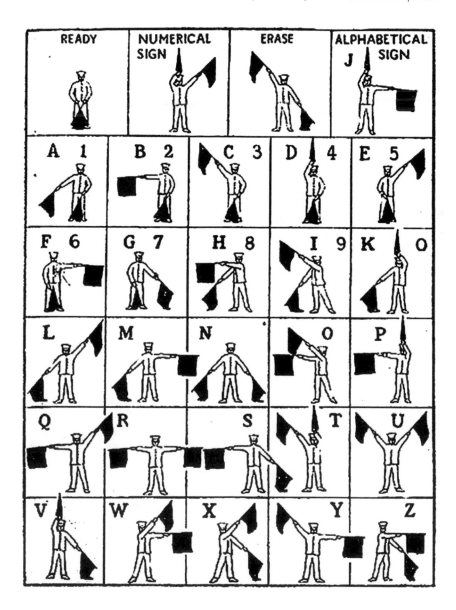

signalling a space, or "ready"). Then, because the flags cannot be told apart, of the remaining possibilities, half are impractical. That leaves just 29 possibilities with static flags—enough to be used for 26 letters plus three special symbols: a 'space' to separate words, an 'erase' option, and a 'switch to numbers', the equivalent of the Num Lock key on your computer keyboard.

So, with 29 possibilities, we calculate that each flag position encodes just under five bits of information.[22] In other words, sending one semaphore symbol is equivalent to choosing one out of a possible 29, which would require at most five yes/no choices if we were using a sequence of Paul Revere–style binary choices.[23]

Early computer scientists adopted a scheme known as the American Standard Code for Information Interchange (ASCII), which used seven bits for encoding the characters in messages. This gave 128 possibilities, which was enough for upper and lower case alphabets, numerals, punctuation, and a set of control characters used for physical operations on teletype machines (some modelled on typewriter functions such as 'tab' and 'carriage return').

These characters could be represented very conveniently in eight-bit groups which became known as bytes. The extra bit was used by different manufacturers for various purposes, for example as a 'parity bit' for error detection.[24] So, under the still widely used ASCII coding scheme, 1 byte of 8 bits is used for encoding one character of text.[25]

Landmark number 8 bits of information can encode 256 different numbers.

Numbers and computer memory

How high can you count using your fingers alone? Without thinking very hard, most people would say ten, counting one number per finger (one digit per digit). But we can change the question slightly and ask, with an information-theoretical frame of mind, how many different signals you could indicate using your fingers, just by allowing each of the ten to be in one of two states, folded or extended.

---

[22] Again, strictly speaking, this is true only if all semaphore positions are equally likely. If some letters are more favoured, as is usually the case, this implies some redundancy in the coding, and the message contains less information than it appears to at first sight.

[23] In Terry Pratchett's Discworld books, a semaphore-like system called 'the clacks' is used to transmit messages between cities. It appears to have used eight shutters that could be open or closed, which means each symbol represents eight bits of information, for a maximum possible 256 states.

[24] A parity bit works like this. If, out of the seven bits used to encode the symbol to be transmitted, there are an even number that are set to 1, then the parity bit should be set to 0; otherwise it is set to 1. That means that taking all eight bits into account, there should always be an even number of bits set to 1. Should you receive a byte with an odd number of bits set to 1, this would indicate a transmission error.

[25] The Unicode encoding scheme, now increasingly prevalent, uses up to 4 bytes, and aims to accommodate all the world's languages and writing systems, and many more symbol sets besides. Unicode's first 128 characters match the ASCII set.

Well, with one hand, you could get 32 signals, from 0 to 31, as follows. Adopt a convention that sticking the thumb out signals 16, the index finger signals 8, the middle finger 4, the ring finger 2, and the pinkie 1. If every finger is outstretched, adding them all up makes 31; if none is outstretched, this signals zero, and there's a way to make every number in between. In fact, this is a five-bit binary code. It can do all that semaphore can, with three symbols to spare.

Now, if we count in this way using both hands, then each hand has 32 independent possibilities, so the total number of combinations is 32^2, or 2^{10}, or 1024 possible signals. And that number, 1024, is the computer world's 'thousand'.

> **Landmark number** 10 bits of information can encode 1024 numbers. (when speaking in binary, the big numbers start at ten bits, or 2^{10}).

> **Rule of thumb**
> - $2^{10} = 1024 \sim 1000 = 10^3$
> - Every ten powers of 2 is equivalent to three powers of 10.

In the 1980s, the early years of personal computers, a very popular microprocessor was the Intel 8086. This used internal memory registers (storage locations) comprising 16 bits. Just as with 10 fingers you can signal 2^{10}, or 1024 different numbers, so with 16 bits you can represent 2^{16}, or 65,536 different numbers. These memory registers were pointers to locations in the computer's memory, and the fact that the pointer used 16 bits meant that only 65,536 different memory locations could be directly pointed to (only 64 kB[26] of memory could be directly addressed at any time).[27]

In time, this became a significant problem in developing software for this generation of computer, a problem that was only alleviated when it became the norm to use 32 bits for memory addressing. The increase to 32 bits meant that 2^{32} addressable memory locations were now available. That made 4 GB (4,294,967,296 locations), which represented a massive and necessary increase.

---

[26] Strictly speaking, 64 kB means 64,000 bytes, since the 'k' signifies 1000. However, it's common usage when discussing computer memory or data transmissions to accept a sloppier usage whereby k signifies 1024. There is an alternative rigorous scheme, which calls 1024 bytes a kibibyte, written 'kiB'. Similar formations exist for MiB, GiB, etc. I'll generally stick with the sloppier common usage in this book.

[27] In fact, the total memory space available was 16 times this, namely 1 MB, but this was achieved through complex swapping algorithms, effectively exchanging blocks of memory to allow programs to make use of more space, and to make it look like 1 MB to programs written for those computers.

However, even this became in turn too limiting. Modern desktop and laptop computers now make use of 64 bits, and this limit permits 2^{64} addressable locations. This comes to 16 exabytes, around a billion times the memory of a current high-end laptop. We'll probably be using 64-bit architectures for some time.

So, you can see how quickly the amount of addressable memory increases as we increase the size of the register used to address memory. As the number of bits in the address quadruples from 16 to 64, the addresses that can be represented go from miserly to enormous.

Making every word count

Information theory has been part of the development of computers from the very beginning—so it's not surprising that measuring information is a relatively easy task, provided it's stored on a computer. But even if not formalised *à la* Shannon, information has been with us throughout history, notably in the form of writing. So let's go back and think about the amount of information contained in books.

Tolstoy's *War and Peace* is the iconic 'very long book'—and so it is, at 544,406 words in English translation. The Penguin Clothbound Classics edition needs 1440 pages for this, placing an average of 378 words on each page.

Herman Melville's *Moby Dick* has 206,000 words. In the Wordsworth Classic paperback edition, this runs to 544 pages, giving an average number of words per page as 380.[28] Charles Dickens's *Tale of Two Cities* has 135,420 words. In a Createspace edition, this gives a page count of 302 pages, for a words-per-page average of 448. George Orwell's *Animal Farm* is a rather slim volume. It has 29,966 words, and a Penguin Modern Classics edition uses 144 pages[29] to hold these words: average = 208 words per page.

Landmark numbers
- *War and Peace* has more or less half a million words.
- *War and Peace* has around 1½ thousand pages.
- A classic novel's words per page: around 400.

---

[28] Incidentally, Melville uses a vocabulary of over 17,000 unique words in *Moby Dick*. This is only slightly smaller than the vocabulary employed in the Bible, which has around 18,000 distinct words. Gibbons's six-volume work, *The Decline and Fall of the Roman Empire*, with a vocabulary of over 43,000, has more distinct words in it than *Roget's Thesaurus*, at around 39,000.

[29] Exactly $^1/_{10}$ of the pages needed for *War and Peace*.

The library with the largest number of books is the United States Library of Congress, with around 24 million books. Printed publications before the year 1500 CE (Johannes Gutenberg first printed with moveable type in 1439 CE) are known as *incunabula*,[30] and the Library of Congress holds 5711 of these. Including non-book items, the Library's total collection is around 161 million items. The British Library has a smaller collection of books, at a mere 14 million, but a larger total collection of around 170 million items.

In 2010, Google undertook to identify how many books had ever been published in the world (counting just one copy of each). They used a strict definition of what counted as a book, using similar rules to those used for allocate ISBNs (International Standard Book Numbers), but excluding things like maps, audio recordings, and T-shirts, which can be allocated ISBNs. They then applied their wizardry to exclude duplicates, and in the end came up with a total of just under 130 million books.

Landmark number Around 130 million books have been published.

Codes and redundancy

Whenever information is recorded, whether it is through the sharpened reed of a Sumerian scribe, by William Shakespeare's pen, or by voice-recognition software taking dictation automatically, it must be encoded. As I type this sentence, the thoughts in my head are being encoded into language, the language is being encoded into letters that I record by pressing keys, and the computer is encoding those keypresses into a form that ultimately finds its way to a storage device somewhere in cyberspace (probably going through several additional encoding steps along the way). Information theory can analyse and quantify each of these coding stages—and one of the important aspects of this coding is redundancy.

Just about all stored information involves some degree of redundancy. Try a game that Claude Shannon invented to investigate how much redundancy there is in written English. Have a friend take a random passage from a novel and ask you to guess the first letter, and then reveal what it is, right or wrong. The chances are that you'll be wrong. But then try to guess the next letter. There's a fair chance that you'll guess right—for example, if the first letter was a

---

[30] The Latin word means 'swaddling clothes' or 'cradle'.

't', a good guess would be 'h'. And so on. You'll find that you can correctly guess a sizeable proportion of the letters.[31]

Shannon's experiments led him to conclude that for hundred-letter sequences of English text there was a redundancy of around 75%. Only a quarter of the letters were really conveying new information, the rest of the letters were, to a degree, predictable and so were only removing a slight amount of uncertainty. This means that in the two steps of coding from thought to language and language to writing, a large amount of redundancy is incorporated into the encoded text.

This is not necessarily a bad thing. Suppose you were faced with the task of restoring a fading, illegible-in-parts manuscript—redundancy is precisely what might help you fill in the gaps. Just as in Shannon's game, clever guessing can help complete the text. The Rosetta Stone used to decipher Egyptian hiero-glyphics was an inscription containing the same message in three languages. Redundancy provided the key to unlocking the reading of hieroglyphics. The Rosetta Disk project, with its multiple versions of the same information in different languages and its policy of making many copies, is relying on redun-dancy to ensure longevity of at least some copies.

These are some of the questions addressed by information theory: how to encode information as efficiently as possible, but also how to introduce redun-dancy in a controlled and optimal way. All modern computer communications include mechanisms for error detection and correction, which might require re-transmission of the message, all of which happens quietly without us being aware of it. We may curse the failings of Internet communications, but they are extraordinarily reliable, and the reason can be traced directly back to Shannon and his fellow pioneers of information theory.

Encoding the Bible

Written English text contains a significant proportion of redundant informa-tion. Can this be measured? The King James Bible contains roughly 783,000 words. How much information is encoded in those words?

A plain-text version of the King James Bible (that's to say, without any fancy formatting) downloaded from the Internet, has a file size of more or less 5,000,000 bytes[32] and at eight bits per byte that makes around 40 million bits.

---

[31] You'll have guessed that this is the principle behind predictive text on a smartphone (for good or for ill).

[32] Quick reasonability check: this means 6.45 bytes per word. Take away 1 byte per word for white space. This gives an average word length of 5.45, which seems entirely plausible.

But as we've seen, English has a high degree of redundancy. The plain-text Bible is not an optimal encoding, and so its file size is not very useful as a measure of information. But we could store the words of the Bible in a more efficient way, and still be able to reconstruct it when needed. For example, we could have a scheme whereby any frequently used words, say 'Jerusalem' with 956 occurrences or 'wherefore' with 458, could be replaced by code numbers, and, provided we also supplied a look-up table that turned those code numbers back to words, this scheme would reduce the number of bytes needed for storage.

This approach is pretty much in the spirit of what computer programs like **WinZip**[33] do: they analyse the file for repetitions and other patterns, and instead of storing the text itself, they store the method for recreating the text. If the zipping process were optimal, it would be able to reduce the text to a file the size of which would equate to the information content of the text. In other words, it would reduce redundancy in the compressed form of the text to zero.

In his book *In the Beginning Was Information*, Dr Werner Gitt calculates that the information content of the King James Bible is approximately 17.6 million bits, which is 44% of the size of the full text. In other words, 56% is a measure of the language redundancy in the text.

Simply using a zip program on the plain-text Bible file yields a .zip file of 2283 kilobytes or around 18.7 million bits, 47% of the full text. Comparing this with Dr Gitt's 44%, it seems that in this case the zipping process is relatively efficient. For comparison, a similar exercise on Herman Melville's *Moby Dick* shows that the zipped version is 0.5 MB versus an unzipped 1.23 MB, giving an information density of 41%, with 59% redundancy.

Landmark numbers
- Words in the Bible—783,000
- Words per page of the Bible—650
- Symbols per page of the Bible—3600
- Information per page of the Bible—1.8kB

Is a picture worth a thousand words?

I tackle this section with some hesitation. As I write this in 2017, I am painfully aware that by the time it is published and read, technology will have made significant advances. Nevertheless, a chapter on the numbers involved in information

---

[33] Other archiving and compression software makes use of the same principles.

technology and data would be incomplete without covering the storage of digital data. Caveats out of the way, here are some typical file sizes for digital media:

- A typical MP3 song: 3.5 MB—more than the entire Bible in zipped plain-text form—around 1 MB per minute of music.
- A 12-megapixel still image: 6.5 MB.
- A movie in DVD format: 4 GB, or around 1000 times as much as an MP3 song.[34]
- A movie in HD format: 12 GB, or about three times the size of a standard DVD movie.
- A movie in '4K'[35] format: 120 GB, or about 10 times the size of the HD movie.

Storage capacity

At the time of writing, it is possible to purchase at reasonable price an external portable hard drive with a capacity of 8 terabytes. For the media files listed above, this will hold:

- Around 70 '4K' movies
- Around 700 HD movies
- Around 2000 DVD movies
- Around 1.2 million still photos
- Around 8 million minutes (15 years' playing time) of MP3 music

Big data centres

So, if it's so easy to find data storage for 8 terabytes, how much data does a company like Google hold? Randal Monroe, of XKCD.com fame, has estimated this as around 10–15 exabytes of storage. But what is an exabyte?

Here's a reminder:

1000 bytes make a kilobyte ('thousand'—10^3)
1000 kB make a megabyte ('million'—10^6)
1000 MB make a gigabyte ('billion'–10^9)

---

[34] Time for a plausibility check: a movie requires 1000 times the information storage compared with a music clip. Well, a movie could be 50 times as long as a music track, with video having 20 times the information density of audio.

[35] 4K is the collective term for a number of video standards with horizontal resolutions of approximately 4000 pixels. By way of contrast, the highest HD 'high definition' standard, called 1440p, has a width of 2560 pixels.

1000 GB make a terabyte ('trillion'—10^{12})

1000 TB make a petabyte ('quadrillion'—10^{15})

1000 PB make an **exabyte** ('quintillion'—10^{18})

1000 EB make a zettabyte ('sextillion'—10^{21})

1000 ZB make a yottabyte ('septillion'—10^{24})

Or, if you prefer (and remembering that 2^{10} is the information world's 'thousand'):

1024 bytes make a kibibyte ('thousand'—2^{10})

1024 kiB make a mebibyte ('million'—2^{20})

1024 MiB make a gibibyte ('billion'—2^{30})

1024 GiB make a tebibyte ('trillion'—2^{40})

1024 TiB make a pebibyte ('quadrillion'—2^{50})

1024 PiB make an **exbibyte** ('quintillion'—2^{60})

1024 EiB make a zebibyte ('sextillion'—2^{70})

1024 ZiB make a yobibyte ('septillion'—2^{80})

So, it's estimated that Google has data storage equivalent to about 1 to 2 million of those 8 TB drives that I can find for sale. A million hard drives—let's picture that as a **visualisation** exercise. Maybe a warehouse-sized space with 200 rows of 250 drives each, all stacked 20 layers deep. That'll do it.

And Big Brother? How much data does he have? Well, we don't truly know, but the US National Security Agency has a huge data centre in Utah. Naturally, the capacity is kept secret. One expert, Brewster Kahle, has taken a **divide and conquer** approach. He estimates from the building size that it could hold 10,000 racks of servers, each rack holding 1.2 petabytes. That gives us 12 exabytes, very much the same scale as the Google facility.

Let me be the first to inform you that...

The number of pages in **Tolstoy's** *War and Peace* is
   10 × the number of pages in **Orwell's** *Animal Farm*.

The storage capacity of the first **Apple iPod Mini** music player (4 GB) was
   1,048,576 × the memory of **the first Apple][™** computer (4 kB)

The **British Library's** collection of items (170 million) is
   340 × the supposed collection of the ancient **Library of Alexandria** (500,000).

The number of frames in *The Hobbit* **trilogy of films** at 48 fps (1,365,000) is
   14 × the number of words in *The Hobbit* book by **Tolkien** (95,400).

Let Me Count the Ways

The Biggest Numbers in the Book

There are many possible realities, infinitely many. Yet most of them are not…alive. Most of them are like books that no one ever actually wrote. A group-mind, like humanity's, lights up one given world. What makes this world different from some ghostly alternative universe is that we actually live here.   **Rudy Rucker**

Which of these is the biggest number?
- ☐ Possible starting hands in Hold 'Em Poker (two cards dealt)
- ☐ Ways a travelling salesman can visit six towns (and return home)
- ☐ Digits used in writing a googol in binary
- ☐ Ways of seating six people around a table

Mathematically big numbers

Although this is a book about practical numeracy and not mathematics, a book about big numbers would not be complete without mentioning the very big numbers that arise in mathematics and especially the area of mathematics called combinatorics.

We've been taking a very rough-and-ready approach to numbers. We've been happy with a 'more or less' approach, and in truth we've been concerned almost entirely with getting to grips with the approximate magnitude of numbers. The question 'Is that a big number?' really has been our primary focus. This is not a book of mathematics. If it had been, we would have investigated quite different properties of the numbers we've discussed, and we would have been much more rigorous in how we described numbers. We'd have been careful to ensure that it was understood that integers are precise whole numbers, that many of their mathematical properties only make sense when they are precise. We would have made clear that real numbers are precise too—even the ones whose decimal representation can never be written out in full. A mathematical book on numbers might cover such wonderful inventions (or are they discoveries?) as transfinite numbers, imaginary numbers, even surreal numbers.

This is a book of practical numeracy, though, and we won't spend too long on these mathematical topics. In fact, we will touch on them only where they have interesting implications for answering the question 'Is that a big number?'

But let's start with the kinds of number that can get much larger than any of the numbers we've seen so far, and which undoubtedly have importance in the real world.

Combinatorics

It's been said of the mathematical topic of combinatorics that it is just counting. Counting is certainly at the heart of the subject. How many different bridge hands of 13 cards can be dealt from a pack of 52? How many different pizzas can you create if there are six optional toppings? This is the sort of question typically used to introduce the subject.

Something you might note is that both of these questions are to do with counting potential states of the world, not actual things. No one is proposing that you make the 64 possible pizzas and then count them one by one; still less that you would deal the 635 billion possible combinations of 13 cards from 52.

We talk about this as 'counting', but these numbers don't arise from actual counting at all. They all come from calculations of combinations and permutations. And the counting in combinatorics is very often done in order to assess probabilities. What's the chance of a bridge hand with all red cards? About 1 in 60,000. What's the chance that a randomly chosen pizza would have pepperoni but no mushrooms among the toppings? One in four. So combinatorics and probability often go hand in hand.

This is an area of study where it's really very common to come across very big numbers. The number of different arrangements that a pack of 52 cards can be shuffled into is roughly 8×10^{67}. That's a stupidly big number. It's bigger, by far, than the diameter of the universe expressed in millimetres. If you were to add the two jokers to the pack, then you could multiply that number by another

2862 to take it well over 10^{71}. These are numbers way bigger than any that you would expect to encounter in the physical world, even in the realms of cosmology. They are numbers that belong in the world of possibilities, not actualities. But sometimes we need to contemplate possibilities.

How hard is that problem?

The computers and smart devices that increasingly serve us in many ways are all simply following sets of instructions, flipping astonishingly many switches very, very rapidly. The instructions they follow are steps in computer programs, and they are written in computer languages. And the idea behind a computer program, the essence of a computer program, is an algorithm.

Suppose we're buying a secondhand car, and have found on the web ten candidates—cars that match our selection criteria. We want to find which of them has the lowest price. It's a very simple problem, I'll admit. Here's an algorithm to do this systematically:

- Note the first car on the list. Call this one the 'best so far' and write down its price and its registration.
- Take the next car on the list. If it's cheaper than one currently designated as the 'best so far', then it becomes the new 'best so far', and you overwrite the price and the registration that you had previously written down.
- Repeat the step above until there are no more on the list.
- Whichever car's details has most recently been written down as 'best so far' at the end of the process will be the cheapest.

That's all an algorithm is. This example, simple though it is, gives the idea: it's just a list of steps to follow. It's not a computer program (because it's not in a computer language), but it is the **idea** behind a program.

Computer scientists, understandably, are concerned with the complexity[36] of algorithms. If you're faced with the task of analysing statistical patterns in a Twitter feed, you want algorithms that get through as many tweets as possible, in as little time as possible. If you can do it in a quarter of the time that your competitors take, you may have a commercial edge over them.

---

[36] 'Complexity' might not mean what you think it means in this case. It's really just a measure of how many steps need to be taken, not how 'complicated' the steps are. And the number of steps is important because it points to how long it might take for an implementation of the algorithm to run on a computer.

In the example of finding the cheapest car, the amount of work that needs to be done is approximately proportional to the size of the list. Working through a list of 20 cars would be twice the effort of working through a list of 10 cars. Working through a list of 1000 cars would take 100 times as long. The computer scientists who study algorithms would characterise this algorithm as having 'of the order of n operations'. They would describe its complexity as $O(n)$, the so-called big-O notation, meaning that as the number of items involved in the search increases, so the amount of work involved increases—and increases proportionally.

If you had the slightly more onerous task of sorting the list of cars into price order, you would need a sorting algorithm. Here's what a simple sorting algorithm might look like:

- Find the cheapest car and put it in position 1.
- Find the cheapest car among those left, and put it in the next position.
- Repeat the above instruction until there are no more in the list.

The number of comparisons to be made in carrying out this algorithm can easily be calculated. To find the cheapest of a list of n cars would require $n - 1$ comparisons. To find the second cheapest needs $n - 2$ comparisons (one has been eliminated), and so on. So, as it turns out, sorting 2 items needs 1 comparison, sorting 3 items needs 3 comparisons, sorting 4 needs 6, and sorting 5 needs 10. It's not hard to show that, using this algorithm, sorting n items needs $(n^2 - n)/2$ comparisons. As the list gets longer, this formula gives answers that get bigger faster and faster. So, while 5 items need 10 comparisons, 10 items need 45 comparisons, and 15 need 105. A thousand items would need almost half a million comparisons. This is the sort of insight we're looking for, because it helps us to choose between different algorithms. When we're looking at the complexity of algorithms, we usually ignore any powers smaller than the biggest (because when the numbers get very big, the highest powers will swamp all others), and so we say this algorithm has $O(n^2)$ time complexity.[37]

Any algorithms with a time complexity based on a power of n (n, n^2, n^3, etc.) are characterised as having 'polynomial complexity'. In the scale of things, this is generally regarded as of middling complexity.

---

[37] This is an inefficient sorting algorithm, but easy to understand. Commonly used sorting algorithms would have complexity $O(n \cdot \log n)$. If the list of items is already almost sorted (for example, if you're adding a few values to an already-sorted column in a spreadsheet), this can be improved further.

The next step up is 'exponential complexity'. Imagine a simple combination lock for a bicycle, with 4 tumblers, each with 10 possible settings. That gives a total of 10,000 possible combinations. Testing them one at a time, taking one second for each combination, would take your bike thief around 2¾ hours. Not great security, but good enough for the purpose. Adding just one more tumbler, though, increases the combinations, and hence the 'solution time' by an order of magnitude, to over 27 hours. Increasing the number of tumblers to 8 would give a lock that would take more than 3 years to run through all the possibilities. This is a problem that has time complexity $O(10^n)$, n being the number of tumblers, and is of exponential-time complexity.

That's not the end of the story: there exist problems of greater complexity than this. The naive or brute-force solution to the Travelling Salesman Problem described below, for example, is of factorial[38] time complexity, and factorials get really big, really quickly.

Computer scientists care about these things because the time complexity measure tells you what problems can and cannot be solved in a practical way and in a realistic timeframe. All our Internet security—all our online banking and buying—is based on the fact that (for the time being) certain algorithms (notably the factoring of numbers) are of sufficient complexity that, if the number used for encoding is big enough, no real-world computer can practically carry out the code-cracking algorithm quickly enough to jeopardise security.

So, this is a whole new way of thinking about big numbers: not just how big they are, but how fast they grow when some parameter increases. Just as the astronomer cares mostly about the scale, the order of magnitude of the numbers that she works with, and not so much about the significand (How many stars in a galaxy? Between 100 billion and 200 billion), so the student of algorithms is more concerned with the order of complexity of his algorithm: it tells him if his algorithm is likely to be efficient, or is doomed to fail when it's applied to large or even modest sizes of input data.

Who'd be a travelling salesman?

In case you missed it: the Travelling Salesman Problem is a problem in the mathematical field of combinatorics. It considers a salesman who has to plan

---

[38] The factorial of an integer is calculated by multiplying together all the integers up to and including that number. So, the factorial of (say) 6, usually written as '6!' is $1 \times 2 \times 3 \times 4 \times 5 \times 6 = 720$. Factorials get very big: 10! is 3,628,800; 20! is more than 2 quintillion.

a driving trip to visit a number of different cities. Naturally, he would like to spend as little time on the road as possible—so he plans a trip that makes one big circuit. The problem is to find the shortest[39] route.

The question is, why is this problem famous, notorious even? The answer has two parts. While it's a very easy problem to state and to understand, and even though we know an exact method of solving it in theory,[40] it's a very hard problem to solve in practice, simply because there are so many possibilities. If there are just three cities (including 'home' as one of them), then the problem is solved, there is only one possibility, namely to travel from H →A →B →H (or the reverse route, which has the same length). For four cities, home and three others, there are only three possible tours (again ignoring trips that are the same, but in the reverse order). For five cities, there are twelve possibilities—so a little more work. But from then on the numbers get big rather quickly:

Cities	Routes
Home + 2	1
Home + 3	3
Home + 4	12
Home + 5	60
Home + 6	360
Home + 7	2520
Home + 8	20,160
Home + 9	181,440
Home + 10	1.814 million
Home + 11	19.96 million
Home + 12	239.5 million
Home + 13	3.1135 billion
Home + 14	43.59 billion
Home + 15	653.84 billion

.
.
.

[39] 'Shortest' could mean the least in terms of distance, time, cost, or any other equivalent measure.
[40] It's very simple. List all possible routes and choose the shortest. The problem is the time taken to list and evaluate all the possibilities.

Cities	Routes
Home + 20	1.216 quintillion
.	.
.	.
.	.
Home + 25	7.756 septillion
.	.
.	.
.	.
Home + 30	132.6 nonillion (132.6 × 10^{30})

In fact, this is growing faster than exponentially:[41] each step is bigger than the previous by a bigger and bigger factor.

We've got big computers, right? These big numbers are child's play for them, not so? Well, not so. When something starts showing more than exponential growth, we know that a point will come, and come quite soon, where the cost of solving the problem **does not scale**. That is, the rate at which the cost/time/effort of computing the solution grows is far faster than the rate at which the scale of the problem grows, by an amount that is unsustainable.

The second reason that this is a big deal is that this problem is (surprise!) not actually about travelling salesmen—that just happens to be a straightforward way of expressing the problem. It's not only of concern to sales departments planning the trips of their on-the-road representatives. In fact, there are plenty of problems that are exactly or approximately equivalent to the travelling salesman problem where good solutions would be really useful.

They arise in many fields: in the drilling of printed circuit boards, in maintaining gas turbine engines, in X-ray crystallography, in genome sequencing, in computer wiring, in product picking in warehouses, in vehicle routing and delivery scheduling, in printed circuit board layout, in cellphone communication, in error-correcting codes, even in game programming where decisions are made as to how to render a three-dimensional scene on the screen.

This makes for a whole family of related problems for which more efficient ways to find solutions would be of huge benefit. These are problems where we know an algorithm for how to arrive at a solution, but the sheer computing

---

[41] Each number here is half the factorial of the number of cities ($n!/2$).

effort of calculating a solution for a large set of inputs is prohibitively difficult. And the problem is simply the big numbers.

What is a number anyway?

In the beginning were the counting numbers: 1, 2, 3 . . . From that point, people have extended the concept of 'number' in more and more 'unnatural' ways. Fractions simply inverted the concept of number. If there was one loaf for four hunter-gatherers, well, the whole had to be broken ('fractured') into fragments or fractions. The concept of zero was a necessary adjunct to the 'Arabic' (really Indian) system of numerals that allowed an open-ended notation for numbers and it may have been used as early as the seventh century CE.

Negative numbers were plainly an absurdity. How could there be a number that was less than nothing, and that, when added to five, gave a result of three? But, in the world of trading and credits and debits, you could see how this backwards arithmetic could work, and work consistently, and be useful. And so negative numbers were admitted into the club . . .[42]

Those qualities, namely consistency and utility, also lie behind the adoption of imaginary numbers and complex numbers. Many people still struggle with these (the name imaginary doesn't help—these numbers are no more imaginary than zero is imaginary—like all numbers, they are mathematical constructs). But being consistent and astonishingly useful makes them highly valued newcomers, becoming widely accepted in the eighteenth century.

And it doesn't end there: Georg Cantor in the nineteenth century found a way of classifying and labelling degrees of infinity, creating the transfinite numbers. Then there are the numbers called p-adic numbers, which can have an infinite number of digits to the left of the decimal separator. Hyperreal and surreal numbers can represent infinitesimal quantities. All these new mathematical objects have to, in effect, pass an admission test to join the numbers club. They are judged for their consistency (insofar as they incorporate the established numbers, they must not violate their established properties) and utility (if they're useful, they will be adopted). The inspirational J. H. Conway wrote a book called *On Numbers and Games*, in which he develops a class of numbers

---

[42] A passing thought: which number is smaller? 0.1 or −1? The story is told of Donald Trump passing a beggar and claiming that the beggar was $9bn richer than he was, Trump being in serious debt at the time. No doubt the beggar took a different view of the matter. This issue, of large negative numbers, is a serious one when trying to construct statements along the lines of 'x% of the world's people own y% of the world's wealth'.

he calls surreal numbers, and then proceeds to identify these numbers with positions in a variety of pen-and-paper games. And so mathematicians keep adding new kinds of objects to what they class as 'numbers'.

What about googol and googolplex?

When Edward Kasner was writing his book *The Mathematical Imagination* (1940), he asked his nine-year-old nephew Milton Sirotta for a name for 10^{100}, a number he was using as an example to show the difference between a very large number and infinity. 'Googol' it was dubbed, and although the number has no specific significance, other than its memorability, it has established itself as a truly landmark number. In physical terms, it is more than the ratio of the mass of the visible universe to the mass of an electron.

Once you have a recipe for creating big numbers, you can make as many as you like, bigger and bigger still. So, the googolplex, coined also by Kasner and his nephew Sirotta, is 10^{googol}. This number could never even be written down in full: the universe doesn't have space to hold it, or materials to do it.

What about Graham's number?

The 1980 edition of the *Guinness Book of World Records* first cited Graham's number as 'the most massive finite number ever used in a serious mathematical proof'. It no longer holds that crown, but it is representative of a class of numbers where even saying what the number is, is problematic. Unsurprisingly, Graham's number arises from a problem in combinatorics.

As we've tackled bigger and bigger numbers, from the small counting numbers, past the comfort zone of a thousand, into the astronomical numbers, we've needed to use different strategies to mentally accommodate them. We've rearranged the way we think about the numbers, to shift and rebalance the conceptual burden. This has sometimes meant a move from a direct understanding of a number to an understanding of the algorithms needed for creating the number.

For example, when we decide that 'quintillion' is no longer very much clearer than 10^{18}, and we switch to using the scientific notation, then we are adopting an algorithm (in this case calculating a power of 10) as our primary way of understanding the number. Graham's number is an extreme version of this shift. Any explanation of it is almost entirely an explanation of the algorithm that would be used to construct it. It's a recipe for taking powers of numbers unimaginably many times. Grasping the algorithm is hard enough. The number itself is beyond conceptualising. No one even knows what the first digit is. Even though the method of constructing the number is well defined, no-one has ever been through the steps—no-one ever could.

At this point, you might ask: is there **any** way of understanding how big these numbers are? And the answer is no, not really, not directly. The only way is by understanding the process by which these numbers are reached, and expressing the steps in these processes using small numbers from our comfort zone.

You might find yourself asking: Is Graham's number even a number at all? That's a very good question. All we know is a method to construct it, but it's a method that could never be followed, as it would require more space and time than the universe provides. But when you think of it, every number we write down, even the £2.50 price of my coffee, is not the number itself, but a method of constructing the number, an algorithm. The difference is just one of the degree of familiarity with the process.

What about infinity?

We know that, even if we could never count them, the counting numbers go on forever. However high we counted, we could never reach a point where we'd say, that's it folks, that's all the counting numbers there are. We don't know if the universe is infinite, but there is no upper limit to the lengths and the masses and the time durations that we can reason and calculate with. But it's one thing to say the range of numbers is without limit, quite another to talk about infinity as a thing in itself. Infinity is not a destination—it has no place on our number line.

Nonetheless, as noted above, in the nineteenth century, Georg Cantor bent his mind to the task of reasoning about numbers beyond the finite, and reasoning in a way that was consistent with established schemes of numbers and that was likely to prove useful. His work inspired strong reaction, as you might expect—after all, he was talking about mind-bending stuff, classifying different kinds of infinities—but was it was in the end accepted and adopted within the mathematical community. Indeed, David Hilbert, one of the leading figures of mathematics in the nineteenth and twentieth centuries, said in praise of Cantor's work: 'No-one shall expel us from the paradise that Cantor has created.'

What Cantor did was to construct a scheme for creating numbers of a new kind—the transfinite numbers—that apply to infinite sets, in the same way as the natural numbers apply to finite sets. Every transfinite number is an 'infinity', and is therefore bigger than any finite number that you could name. So, to ask 'Is that a big number?' of one of Cantor's transfinite numbers doesn't really even make sense. It's in a realm of its own.

You just have to work the numbers...

The quantity of real numbers between 0 and 1 (the 'continuum' transfinite number) is
  the same as the quantity of all the real numbers there are.

The number of ways to rearrange a list of 70 items is
  a little more than a googol.

In his work *The Sand Reckoner*, Archimedes estimated the number of grains of sand that would fill the universe as
  8×10^{63}.

The quantity of all the rational numbers there are (the 'aleph-null' transfinite number) is
  the same as the quantity of all the natural numbers there are.

PART 4

Numbers in Public Life

The Numerate Citizen

The good life is one inspired by love and guided by knowledge.   **Bertrand Russell[1]**

Every day, we make decisions that affect not only our own lives but also those of the people around us. Some choices have immediate and direct consequences. An example: the votes we cast. Some choices are silent: the products we buy and the websites we visit are tracked and can influence business decisions made far away. Some choices are noisier: the arguments we express publicly. Some people raise their voices and play active roles in public life, others touch and influence others more gently in their roles as parents, teachers, or work colleagues. Quietly or noisily, our choices make ripples, and collectively these choices shape our society. We should make those choices, as best we can, guided by knowledge.

The bones of the world

Leonardo da Vinci dissected human and animal bodies to understand the muscles and the skeletons of the people and animals he drew and painted. He understood that surface appearances, which could be altered and distorted through choice or accident, were not enough. For Leonardo, the artist, the musculature gave him the contours, the bulk, the power and movement in the figures he depicted. But deeper than muscles, the bones, and the way they connected together, showed him the range of articulation possible for the human body. This is one of the factors that give his images such great energy. But to get to that understanding, Leonardo had to look beneath the skin.

[1] Russell, Bertrand, *What I Believe*, Routledge, 2013, p.10. © The Bertrand Russell Peace Foundation Ltd. Permission granted by the Publishers Taylor & Francis.

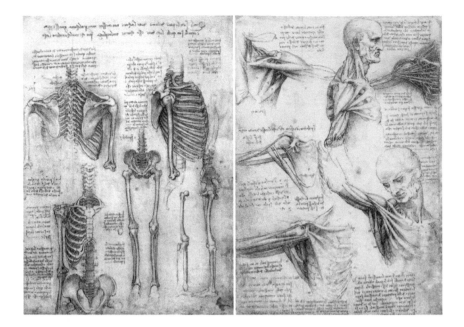

We all struggle to understand the world. All that we see directly are surfaces, and they can be confusing. Truly, there are too many facts, too many numbers, for us properly to absorb, interpret, and assess everything we are told.

It's too easy to fall prey to confirmation bias. We match new information to our preconceptions and filter accordingly. If the news meshes easily with our worldview, we accept, endorse, and share it, uncritically. Where information threatens our prejudices, we seek a reason to reject it, seizing on any easy rationalisation.

We can't think through every decision from scratch. Our choices are always made by reference to an internalised model of the world. If that model matches reality sufficiently closely, the choices we make have a better chance of turning out well. And a model based on surface appearances will likely be a poor model. Better to follow the anatomical example of Leonardo, and try to look for what lies below the surface. And that means understanding the muscles and bones of the world. If our values and beliefs are connected to a deep understanding, they will be not only more persuasive, and less vulnerable to emotive manipulation, but also more flexible when there is evidence that change is needed.

This may sound awfully naive and unsophisticated, but I know of no way of creating such a mental model other than by basing it on a numerate, logical, and rational understanding of the world we live in. If we honestly seek true views of the world, then numbers will support us when we are right, and challenge us when we are wrong.

That's not to say there is one, absolute, 'right' point of view. Scientific understanding changes, and science proceeds by finding explanations that are progressively 'less wrong'. The same set of facts may often support different interpretations of cause and consequence. People hold different values, and prioritise different goals. When views and values clash in public debate, numbers are frequently used as weapons, either as part of genuine debate or very often simply as distractions. Understanding where numbers come from and what they mean can help to make sense of these battles and allow us to judge whether the arguments relate to superficialities or deeper reality. Skin or bones?

Global numbers

Increasingly, people throughout the world face the same problems. Not just similar problems, but problems that cross national boundaries, and that national governments cannot address on their own. Pollution of the soil, seas, and skies; climate change; loss of habitats and biodiversity; the regional and global impacts of dysfunctional states; tax-regime arbitrage by multinational corporations; organised crime; and, yes, terrorism. These problems, highly complex and difficult in themselves, are nearly impossible to deal with given their 'international' nature.

As private individuals, it is hard to see that we can have any influence on these matters. As citizens of our respective nations, we can seek to elect governments that place what we feel to be an appropriate weight upon these wider issues. As citizens of the world, we can raise our voices where we feel our voices will make positive contributions. We can donate time, effort, and money to campaigns we believe in.

But how will we know what 'appropriate weight' means? How will we know where our energies and resources are best expended? As Bertrand Russell suggests, the love in our hearts may inspire us, but for sound guidance, we must rely on the content of our heads, on knowledge. And knowledge means numbers.

This part of the book addresses several topics of national and global importance where understanding the numbers makes a difference. It touches on money and economics; growth in the population of humans and domestic animals; decline in the populations of wild animals; measurement of variation and inequality; attempts to place numbers on quality of life.

It is by no means an exhaustive list. Rather, it is an attempt to show, using a selection of examples, how a numerate approach can throw light on, and establish context for understanding, some very important issues.

And the first of these issues is money.

Who Wants to be a Millionaire?

Counting the Cash

A billion here, a billion there, and pretty soon you're talking about real money.

Everett Dirksen

If you can actually count your money, then you're not a rich man.   J. Paul Getty

---

Which of these is the greatest amount?
☐ Cost of *Apollo* moon programme (in 2016 dollars)
☐ GDP of Kuwait in 2016
☐ Turnover of Apple in 2016
☐ Value of Russia's gold reserves (July 2016)

Tallysticks and stockholders

Commerce has always depended on debt. It's not always feasible or convenient to complete commercial exchanges in full the instant a deal is struck. Modern accounting systems can reliably record credits and debits. But long before the invention of double-entry bookkeeping, a scheme was developed in mediaeval Europe by which debts could be recorded in a tamper-proof way.

A length of wood was marked with the amount of the debt, by making notches on one or more edges. The tallystick, as it was called, was then split down the middle in such a way that the two pieces carried matching notches. The way it was cut, one piece was longer. This was called the stock, and was retained by the lender, the stockholder. The shorter piece was called the foil.

The amount of money involved was recorded as a series of notches of varying widths. According to the *Dialogue Concerning the Exchequer*, a twelfth-century treatise by Richard FitzNeal:

The manner of cutting is as follows. At the top of the tally a cut is made, the thickness of the palm of the hand, to represent a thousand pounds; then a hundred pounds by a cut the breadth of a thumb; twenty pounds, the breadth of the little finger; a single pound, the width of a swollen barleycorn; a shilling rather narrower; then a penny is marked by a single cut without removing any wood.

When the debt was settled, the stock and the foil were brought together. Crucially, the person settling the debt could be sure that they were making payment to a counterparty who was entitled to that repayment. Matching grain patterns proved that foil and stock had each come from the same original tallystick, and even if the debt had been sold on, the physical stock would have been passed on also, as tangible evidence. Matching notches would prove that the amounts had not been altered. So tallysticks relied on some of the principles that would later make paper money and other financial instruments viable, among them the ideas that tokens of debt could be passed on from one creditor to another, and that money could take the form of information, a number backed by a means of authentication. And now in the twenty-first century, these same principles underpin one of the newest forms of money, the Bitcoin.

Though most money today is held as secure information, through most of history money took the form of rare or valuable physical items, and the most important of those were coins.

What's that in Old Money?

In 1516 a silver mine started operations in Joachimstal, in Bohemia, in what is now the Czech Republic. Silver from the mine was first minted into coins in 1518, and these became known as *Joachimsthalers*. This name was shortened to *thaler* (and later lengthened in a Dutch rendition of the word, to *leeuwendaalder*[2]). These words give us the names of many different units of currency: from the Slovenian *tolar*, the Romanian and Moldovan *leu*, and the Bulgarian *lev*, to of course the American *dollar*.

---

[2] The name references the lions ('leeuwen') on the Dutch version of the coin.

But the Bohemian mint had much earlier precedents. The earliest coins that have been found were in parts of today's Turkey that would have formed part of the ancient kingdom of Lydia, and the oldest is believed to date to around 700 BCE. They are irregular lumps of electrum (a mixture of gold and silver) with a design on one side. What allows us to recognise them as coinage is that they come in standardised weights.

In the third century BCE, the Romans were producing coins at the temple of Juno Moneta. The name Moneta comes from the Latin *monere*, meaning 'to remind, warn, or instruct', and through the name of that temple we find the root of both the words 'mint' and 'money' (as well as 'admonish'). The Romans spread their coinage throughout their empire, the *denarius* being the base unit. 12 *denarii* made a *solidus*, 20 *solidi* made a *libra*.[3]

Charlemagne introduced to Europe a monetary system based on that of the Romans. A pound in money terms was originally a pound weight of silver pennies, *deniers*[4] in French (240 of them) and a shilling was a twentieth part of a pound. Systems modelled on Charlemagne's silver standard became widespread in Europe, and the names of coins such as the *dinero* in Spain reflect this. The system lingered into the 1970s in the UK and those countries that inherited their monetary system from the British Empire.

Where most of Europe was on Charlemagne's silver standard, the Arab and Byzantine worlds were on a gold standard. The Islamic Umayyad dynasty issued gold *dinars* (again from the Latin *denarius*), and the Byzantines issued what became known as *bezants*. To this day, the heraldry term for a yellow roundel is a bezant.

The British unit, the guinea, was a curiosity: a pound plus a shilling. A deal brokered by a middleman could involve a payment by the purchaser of a certain number of guineas, and a payment to the seller, of the same number of pounds, the shillings that made up the difference being pocketed by the middleman as a commission, which was therefore equivalent to 5%. A further curiosity: the guinea, being 21 shillings, was also evenly divisible by 7 and 9 without needing fractions of pennies.

---

[3] In Latin, a pound weight was *libra pondo*. From this, we get *lire, livre, pound*, and even *peso* and *peseta* (literally, 'weighed').

[4] And 'denier', the measure of the fineness of thread, as used in silk stockings, comes from the same origin. The measure in denier, according to the modern definition, is the weight in grams of a 9000 m length of the yarn; so the fineness could be measured by weighing a known length of the fibre. A denier coin, as minted for Charlemagne, would have weighed around 1.2 g.

The word 'cash' has a curious double etymology. The primary derivation is from Middle French *caisse*, from Latin *capsa* (meaning a box, often cylindrical[5]). On the other hand, the earliest Chinese coins, those ones with the square holes and commonly made from copper, were also called 'cash', the word deriving ultimately from the Sanskrit word *karsha*. They were in use in China from the second century BCE and were strung together through the central holes to form 'strings of cash', nominally 1000 to a string, sometimes divided into hundreds. These strings, which were typically carried over the shoulder, were often a few short of the full 1000, since a few were taken as commission in the process of assembling the string.

In the Tang dynasty (more or less 600–900 CE), 'flying cash' came into use. Effectively the first banknotes, these were certificates deemed equivalent to the strings of cash coins and sometimes even bearing a depiction of the string of cash they represented. Also in the Tang dynasty, bolts of silk became a recognised means of settling debts, and thus a form of money. The standardised bolt was 12 metres long and 54 centimetres wide, and was used for high-value transactions.

The first Indian coins date back to the sixth century BCE. The name *rupee* comes from a Hindi word meaning 'shaped' or 'stamped' and has been in use at least since the fourth century BCE, referring to silver coins. Until decimalisation in 1957, a rupee was divided into 16 *annas*, and each anna was divided into 4 *paisas* (the root of the word means 'quarter'), and each *paisa* into 3 *pies*. Upon decimalisation, the rupee became equal to 100 (new) paisas.

The shilling is still the name of the base currency in Kenya, Uganda, and Tanzania, and, until adoption of the euro in 2002, the *schilling* was Austria's unit of currency. The name derives from Old Norse, meaning a division.

The name 'crown' lives on in the Scandinavian countries' *krone/krona* and the Czech *koruna*. The English silver crown was a coin roughly equivalent to the Spanish dollar, equal to one-quarter of a pound, and, since under the fixed exchange rate system of the early twentieth century a pound was worth four dollars, in the 1940s the crown acquired the nickname 'dollar' in Britain.

Prior to the American Revolution, the name *dollar* was already widely known as a name for the Spanish eight-real piece, which was in common use in the Americas. When a new currency was needed for a new country, this name was adopted in 1775.[6] The eight-real piece could in fact be physically cut into eighths.[7] Two of these one-eighth pieces made a quarter-dollar, which is why the US

---

[5] A small one of which was naturally a 'capsule'.

[6] Spanish dollars were legal tender in the USA until 1857.

[7] This coin gave Captain Flint, the parrot in *Treasure Island* his catchphrase: 'pieces of eight!'

quarter is sometimes referred to as 'two bits'. The word *dime* (the original spelling was the French-style *disme*) derives from the Latin *decima* (one-tenth), and a *nickel* was, well, made from the metal nickel. And a dollar is sometimes a *buck* because, yes, deerskins were used for trading in eighteenth-century America.

Measuring money

Money is used to measure economic muscle in two different ways: a monetary amount may represent **capital**, an accumulated stock of assets, or liabilities—that's money standing still. Or it may represent **income**, flows of revenue, or outgoings—that's money in motion.[8] Confusion between these two uses is widespread (for example, the national debt is a capital amount, an accumulation, while the national deficit is an income amount, a flow). It's the difference between a reservoir and a river.

An amount of money is really just a fancy form of counting, rather than a way of measuring any real physical quantity. An amount of dollars and cents is simply a way of expressing a count in cents. Similarly, for other currencies, an amount of money is just a count of units of the smallest token of legal tender.

When we measure distance, or mass, or time, we can rely on a well-established and definitive system of standard units, accepted globally, the International System of Units (SI). We have no such absolute standard with money, but must choose a currency to suit our purposes. The US dollar, as the pre-eminent currency in today's world, is what we will use to measure money in this book.[9] But even the mighty dollar is far from being a fixed reference point.

Currencies: money's wobbly yardsticks

The exchange rates between currencies vary continually—so, if you're looking for landmarks and hoping to memorise conversion factors, you'll need to refresh your information from time to time. News media are relatively good at converting reported currencies from foreign to domestic currencies: nonetheless,

---

[8] What's the original meaning of 'currency'? Here's the Online Etymology Dictionary entry: *'currency' 1650s, 'condition of flowing,' from Latin currens, present participle of currere 'to run' (see current (adj.)); the sense of a flow or course extended 1699 (by John Locke) to 'circulation of money.'*

[9] As I write this chapter, the UK pound has devalued markedly against the dollar and the euro, the consequence of market sentiment turning against British investments, and the dollar itself has been reducing in value. You, dear reader, will know better than I how that turns out.

it can be very helpful to have some idea of the rough equivalencies between major world currencies. Remember that the absolute values of currencies are less important than the ratios between them, and the way those ratios change with time.

With that in mind, here are some landmark exchange rates valid in late 2017:

Landmark numbers

100 US dollars will buy you more or less

- 125 Australian dollars (AUD)
- 120 Canadian dollars (CAD)
- 95 Swiss francs (CHF)
- 85 Euros (EUR)
- 75 British pounds (GBP)
- 780 Hong Kong dollars (HKD)
- 10,800 Japanese yen (JPY)
- 650 Chinese renminbi (CNY)
- 6400 Indian rupees (INR)
- 5750 Russian roubles (RUB)

Interest and inflation

It's not always easy to answer the question 'Is that a big number?' when asked about an interest rate or an inflation rate. What seems a modest, even small, number can quietly result in large financial effects. Take, for example, a loan of $1000 from a loan shark for the low, low rate of 10% per month (or, rather, don't take such a loan—we'll see why in a minute). Sure, that rate sounds a little higher than you'd expect to pay for a bank loan, but really, it's not crazily high. Is it?

In fact, if you were to neglect to pay the monthly interest, but allowed it to mount up month by month, then after a year you would have accumulated a debt amounting to a whopping $3138.[10] If that same debt of $1000 were left unpaid for just two years, the total accumulated debt would be $9850.

What may seem small differences in interest rates can have large effects. If the rate had been lower, say 5% per month instead of 10%, the total after one year would be 'just' $1796 and after two years $3225. Halving the interest rate reduces the interest paid by much more than half.

---

[10] Calculated as $1000 \times (1.10)^{12}$.

The arithmetic of compound interest is all about working through how growth rates translate into the effects of that growth. It's essentially the same as the logic behind **log scales**. On a log scale, each step represents a consistent factor of increase, a consistent amount of growth. When you have an amount of money (or debt) growing with compound interest, the rate of interest represents the step size you take each period, and the number of periods (months in our example) represents how many steps you take. And we know already that on a log scale we can reach very large values very rapidly.

And the opposite of growth is shrinkage, which is what happens under inflation. And again, seemingly innocuous numbers can have shocking effects when they apply consistently over an extended period.

Remember when a dollar was still a dollar?

Inflation is another complication that stands in the way of a straightforward understanding of monetary numbers. Unlike mass, where there is a fixed and stable reference kilogram, or time or distance, where there are scientifically defined absolute standards, there is no absolute standard for money. Even the gold and silver standards of the past were not absolute. Setting aside exchange rate variation, the value of any currency varies constantly with time, usually devaluing as the force of inflation has its effect. When it comes to money, our mental grasp on the numbers involved is always slipping away.

As with interest, rates of inflation that seem modest enough when expressed as percentage rates can, through the process of compounding, cause surprisingly large effects. The average rate of inflation in the USA in the 1970s was 7.25% per annum (pa). That rate of inflation, 7.25% pa, doesn't seem so brutal—not when compared with tales of hyperinflation from prewar Germany or more recently from Zimbabwe in the 1990s—but it's damaging enough. If you compound that inflation every year for ten years, you end up with a decline in the purchasing power of the dollar of just over 50%. So, a 1980 dollar would have bought half as much as a 1970 dollar. Any money you had stashed under the mattress would have halved in value.

By contrast, in the decade of the 2000s, the US average inflation rate was 2.54% pa, and that meant that the devaluation factor over that decade was only 22%. A 22% loss of value is still a powerful argument against keeping money under the mattress, but it's nowhere near as bad as the situation in the seventies.

Taking a longer view, over the past century, the average US inflation rate has been 3.14% pa. Compounded over 100 years, this represents a loss of value of 95.5%. Put another way, a dollar today buys what 4½ cents would have bought 100 years ago. In the UK, the rate has been on average 4.48%, which compounded gives a loss of value of 98.75%—a pound is worth $^1/_{80}$ of what it was then: 3d in old money, equivalent to 1¼ modern pennies. That difference of 1.34% pa between UK and US inflation, which seems small enough, has, over the course of a century, made the pound worth barely a quarter of what it was relative to the dollar.

Rule of thumb To work out roughly how long an interest rate will take to double your money, divide the number 72 by the interest rate in percent. So, for an interest rate of 6%, it will take 12 years. This is the 'rule of 72'.
Double-check:
$1000 × (1.06)^{12} = $2012
The same calculation will work to tell you how many years it will take an inflation rate to halve the value of your money. So in the 1970s in the USA, the value of the dollar was halved in around 10 years ($^{72}/_{7.25} = 9.9$ years).
The same rule works for any other rates of growth. A website that is attracting 10% more visitors every week will double its audience in $^{72}/_{10} =$ just over 7 weeks.

To compare amounts of money from different years in the past, we need to allow for inflation. We need to qualify the monetary unit we're using by specifying the year it applied—so we talk about 1970 dollars, or 2017 pounds. To compare (say) the cost of the Vietnam war with the cost of the campaign in Iraq, the Vietnam costs must be revalued to a comparable date. (The comparison works out to an estimated $778 billion for Vietnam and $826 billion for Iraq when adjusted for inflation to 2016 dollars.)

Another more subtle effect of inflation is simply that, when it comes to measuring monetary amounts, the accumulated effect of inflation over centuries has resulted in us using for everyday purposes numbers that are very large in themselves. So, an annual salary may be measured in the tens of thousands of dollars or even more, which is already, in the terms of this book, a big number. National budgets are in the billions and even trillions. When it comes to money, to ask 'Is that a big number?' is to ask an even trickier question than usual.

In Jane Austen's *Pride and Prejudice*, the romantic hero Mr Darcy is said to have an annual income of £10,000. Is that a big number? It certainly wouldn't be regarded as such today, but how has the value of the pound changed since 1810?

Inflation makes it especially difficult to understand the meaning of the amounts of money mentioned in books that are set in the past. For your convenience, here's a ready reckoner showing the factor that you should apply to historic amounts of money (pounds or dollars) to bring them to modern equivalents (2016 values):[11]

If a book's set in the year...	That's how long ago?	Multiply UK pounds by...	Multiply US dollars by
2015	1 year	1.02	1.01
2011	5 years	1.12	1.07
2006	10 years	1.33	1.19
2001	15 years	1.52	1.36
1996	20 years	1.7	1.5
1991	25 years	2.0	1.8
1986	30 years	2.7	2.2
1976	40 years	6.6	4.2
1966	50 years	17.1	7.4
1941	75 years	46.3	16.3
1916	100 years	80	22
1866	150 years	109	~
1816	200 years	89	~
1810	**206 years**	**72[12]**	~
1766	250 years	160	~
1716	300 years	189	~
1616	400 years	233	~
1516	500 years	922	~

The factor for 1810, to be applied to Mr Darcy's income, would be 72, meaning that his annual income of £10,000 would, in today's terms, be a whopping

---

[11] As I write, the UK pound is slumping. These figures are pre-Brexit...

[12] It's interesting to note that this figure is lower than for the rows above. This reflects net **defla**tion in the period 1816–1866. From 1819 to 1822, inflation in Britain was negative every year, and reached as low as −13.5% in 1822.

£720,000, which is certainly a big number, and enough to put him among the '1%' of top earners quite easily.

Unconventional indices of purchasing power

Exchange rates don't fully reflect the relative worth of money in different countries. Different goods and services will be subject to their own supply and demand depending on local conditions, and a host of other factors may affect how much your money can actually buy in different countries.

As a semi-humorous way of illustrating this point, in 1986, *The Economist* magazine introduced their Big Mac Index. The idea is that by tracking the price of a McDonald's Big Mac burger (a commodity they reckoned was widely available internationally in a more or less standardised form), the real relative purchasing powers of different currencies could be tracked. The Index is still in operation, 30 years later. As of September 2017, a Big Mac is most expensive in Switzerland ($6.74) and least expensive in the Ukraine ($1.70). In the USA it is $5.30, and in the UK $3.19.

Because McDonald's does not have as strong a presence in Africa as KFC (Kentucky Fried Chicken, as was), the KFC Index was introduced as an alternative measure of purchasing power parity for use in Africa.

In the same vein the Bloomberg business information organisation published a Billy Index, based on the ubiquitous Billy Bookcase[13] from Ikea, the global furniture supplier. As of October 2015, the Billy was cheapest in Slovakia ($39.35), and most costly in Egypt ($101.55). In the USA it was $70, and in the UK $53.

Measuring economies

Money is used to measure many things, not least the size of a country's economy. Here are some Gross Domestic Products[14] (GDP) for various countries in 2016, arranged as a number ladder. As you might expect, all the smallest economies are island states. And remember, every three rungs on this ladder represents a ten-fold increase; so this is a kind of **log scale**: the numbers get much bigger, very fast.

---

[13] Yes, I have one in my home.
[14] The next chapter will cover GDP in more depth. For now, think of it as how much a country earns in a year. It's the national equivalent of annual salary.

$10 million   GDP of Niue = $10 million

$20 million   GDP of Saint Helena, Ascension, and Tristan da Cunha = $18 million

$50 million   GDP of Montserrat = $44 million

$100 million   GDP of Nauru = $150 million

$200 million   GDP of Kiribati = $210 million

$500 million   GDP of British Virgin Islands = $500 million

$1 billion   GDP of Samoa = $1.05 billion

$2 billion   GDP of San Marino = $2.02 billion

$5 billion   GDP of Timor-Leste = $4.98 billion

$10 billion   GDP of Montenegro = $10.6 billion

$20 billion   GDP of Niger = $20.3 billion

$50 billion   GDP of Latvia = $50.9 billion

$100 billion   GDP of Serbia = $101 billion

$200 billion   GDP of Uzbekistan = $202 billion

$500 billion   GDP of Sweden = $498 billion

$1 trillion   GDP of Poland = $1.05 trillion

$2 trillion   GDP of South Korea = $1.93 trillion

$5 trillion   GDP of Japan = $4.93 trillion

$10 trillion   GDP of India = $9.72 trillion

$20 trillion   GDP of USA = $17.42 trillion
GDP of China = $21.3 trillion in terms of purchasing power, or $11.2 trillion at nominal exchange rates[15]

Landmark GDP numbers
- Samoa—$1 billion
- Montenegro—$10 billion
- Serbia—$100 billion
- Poland—$1 trillion
- India—$10 trillion
- USA—$17 trillion
- China—$23 trillion

---

[15] The $11 trillion figure is based on an artificially low exchange rate; the $21 trillion figure is an estimate based on purchasing power in the country.

That's rich!

The **GDP of Germany** (2016: $3.98 trillion) is
  4 × **GDP of Pakistan** (2016: $988 billion)

The **GDP of the UK** (2016: $2.79 trillion) is
  a little larger than **GDP of France** (2016: $2.74 trillion)

The turnover of **PetroChina** (2015: $368 billion) is
  4 × turnover of **Nestlé** (2015: $92.2 billion)

The total taxation in **France** (2016: $1.29 trillion) is
  10 × total taxation in **Finland** (2016: $128 billion)

The turnover of **Volkswagen** (2015: $310 billion) is
  2 × turnover of **General Motors** (2015: $156 billion)

GDP of **Brazil** ($3.14 trillion) is
  40 × GDP of **Bolivia** ($78.3 billion)

Exports from the UK (2016: $800 billion) are
  2 × **exports from Ireland** (2016: $402 billion)

Cost of making *Star Wars: The Force Awakens* ($200 million) was
  1000 × the cost of a Lamborghini Gallardo car ($200,000)

The Bluffer's Guide to National Finances

The tax which will be paid for education is not more than the thousandth part of what will be paid if we leave the people in ignorance.   **Thomas Jefferson**

If you think education is expensive, wait until you see how much ignorance costs.

Barack Obama

How many of us, economists apart, really understand national finances? Who has a good feel for the numbers when it comes to the money that our governments collect and spend on our behalf? Few of us celebrate the tax we relinquish, and few of us properly appreciate the spending that the government makes on our behalf. It seems that every opposition party there's ever been has claimed that they can eliminate inefficiency in government, and, inevitably, incoming regimes transform those promised 'efficiency savings' into 'cuts' on taking power.

Cynicism aside, it is helpful to understand, at least broadly, how government finances work, and the scale of the numbers involved. This is not an economics lesson, but, in the spirit of this book, I hope to paint for you a picture that is accurate enough to allow you to form reliable impressions of the sizes of the numbers involved, but still to keep things simple enough, and not to get snarled up in the many complicating factors that can make economics such a difficult subject to grasp.

These pages use very broad brushstrokes intended to establish primarily the relative scale of the amounts involved, for example that the USA and China have economies that are roughly of the same order of magnitude and that Britain's is a bit more than a tenth of that size.

To include figures from every country would turn this chapter into no more than a series of large spreadsheets. Instead, I have chosen three economies to illustrate these points: the UK, where I live, and the USA and China—as the two largest economies in the world, operating under contrasting circumstances and regimes.

Another problem is that these numbers change over time. The effects of inflation and growth, not to mention dozens of other forces of change, operate to make any numbers that I quote here instantly out of date. The point of this chapter though, is to show **relative** sizes and magnitudes—so I have fixed on a recent year, 2016, where there is a good chance that all the accounting and adjustments have now been settled. Finally, as a standardising currency, US dollars are used.

Having given my caveats, the website IsThatABigNumber.com has pages relating to national finances that will allow you to choose from dozens of countries, to explore numbers like these for the country you choose.

How much do we earn?

'Gross Domestic Product' or GDP: that dry name is our starting point. Just as in a household, the main driver of economic choice is how much money is coming in, so all national economic choices are driven by how much the country is earning.[16]

Let's unpack the terminology: 'Product' refers to how much is produced; 'Domestic' means that it relates to economic activity in the country concerned (as opposed to economic activity in foreign countries that may be owned by citizens); 'Gross' signifies that the number makes no allowance for depreciation or other loss of value. GDP is by no means a perfect way of measuring the economic activity of a country (it doesn't, for example, measure unpaid work or the economic activity of black markets), but it's widely available, and it's generally carefully calculated.

Because we like to express big numbers as **per capita ratios**, we're also very interested in population. Countries vary hugely in size, and we can draw few useful conclusions by comparing only the absolute economic statistics. Using the population of the country as a base allows us to make per capita comparisons, which will have more meaning.

[16] Gross World Product (GWP) is the total of Gross Domestic Product for the whole world.

So:

Measure	UK	USA	China
GDP	$2.79 trillion[17]	$18.56 trillion	$21.27 trillion[18]
			($11.22 trillion officially)
Population	64 million	324 million	1374 million
GDP per capita	$43,300	$57,300	$15,490

Landmark numbers (with considerable approximation, bearing in mind change over time, and imprecision of measurement)
- UK GDP—$2.5 trillion
- US GDP—$20 trillion
- Chinese GDP—$20 trillion

How much does the government take in tax?

Taxes come in many shapes and forms—so these figures attempt to incorporate all forms of taxation.

Measure	UK	USA	China
Tax take	$1.00 trillion	$4.36 trillion[19]	$4.68 trillion[20]
Tax % of GDP	36%	23.5%	22%
Tax per capita	$15,460	$13,460	$3400

---

[17] Numbers for 2016 taken from *CIA World Factbook 2017*. Exchange rates as at 31 December 2016.

[18] When it comes to economics, things are seldom simple. At the time of writing, the Chinese government applies an artificial exchange rate. This means that the purchasing power of money in China is understated if we use the official exchange rate. The bigger number here is the more meaningful equivalent.

[19] Includes Federal, State, and Local Taxes, and 'social contributions'.

[20] For comparison internationally, and for consistency with the Chinese GDP quoted earlier, this (and other figures for China) has been scaled using the same Purchasing Power factor as the GDP.

How much does the government spend?

The government collects tax (and borrows money) in order to fund its spending programme. How much does it spend?

Measure	UK	USA	China
Government spending	$1.10 trillion	$5.66 trillion[21]	$5.50 trillion
Spending % of GDP	39%	30.5%	26%
Spending per capita	$17,200	$17,470	$4000

Landmark numbers (again with considerable approximation)

- UK government spending—$1 trillion
- US government spending—$5.5 trillion
- Chinese government spending—$5.5 trillion, about the same as US spending, for over four times the population.

What's the balance?

If the government spends more than it collects in taxes, there will be a **deficit** for the year, and the government will have to borrow to fund this deficit. Conversely, if it spends less than it collects in taxes, there will be a **surplus**, and the government can repay borrowings, or even (rarely) accumulate an overall surplus.[22] The numbers below are negative, reflecting that in all cases, they refer to deficits.

Measure	UK	USA	China
Surplus/deficit	−$0.10 trillion	−$1.3 trillion[23]	−$0.43 trillion
As % of GDP	−3.6%	−7.0%	−2.0%

[21] Including social benefits.

[22] Surpluses do sometimes arise: In Norway, for example, there was a surplus of $11 billion in 2016.

[23] Including social contributions and social benefits. If we exclude those, the deficit was a much smaller $530 billion, 2.9% of GDP and $1636 per capita.

Measure	UK	USA	China
Per capita	−$1560	−$4000	−$310
As % of spending	−9%	−23%	−15%
Spending/earning	110%	130%	117%

I'm always wary of numbers, like the surplus/deficit, that represent the difference between two large numbers that almost balance. Such numbers can be very sensitive to movements in either of the underlying figures, and can make comparisons quite difficult to interpret. The final line of the table shows an alternative, a ratio of spending to earning for the various countries, which is more easily comparable.

Note also that the deficit is an annual figure, representing a **flow** of money, not an **accumulated amount** of money. It's how much the government is out of pocket for the year concerned.

How much do we owe overall?

The deficit is how much we're out of pocket for a single year, and will have to borrow in that year. The **national debt** is the overall net amount the country has borrowed over the years (and on which it has to pay interest).

Measure	UK	USA	China
National debt	$2.57 trillion	$13.7 trillion	$4.2 trillion
As % of GDP	92%	74%	38%
Per capita	$40,200	$42,300	$5900

How much does that cost us?

The national debt (ND in the table below) is secured by the ability of a government to tax its citizens. Because they have that right, which establishes a reliable source of income, institutions like banks and pension funds and other investors are willing to lend to governments at relatively low interest rates:

they assess that the risks of failing to repay the debt (or to meet interest payments) are low.

Measure	UK	USA	China
Interest on ND	$40 billion[24]	$233 billion	$270 billion[25]
As % of ND	1.6%	1.7%	3.3%
As % of GDP	1.4%	1.3%	1.27%
As % of tax	4.0%	5.3%	5.8%
Per capita	$625	$720	$195

Numbers of this sort are the ones that lead to the heated political discussions around 'austerity'. It raises some hackles that (to take an example from the table) about one dollar in every 20 that is raised by the US Government in taxes goes to pay the interest on the national debt. This book takes no position on this: the sole intention here is to give you some idea of what the numbers mean, and the scale of their magnitude.

We tend to have negative associations with borrowing, and by and large this is justified. Debts are indeed negative wealth. But debts are also empowering: a startup entrepreneur might regard securing a line of credit as a very positive development. A young couple might be thrilled to have a mortgage approved so that they can buy their first home. In the same way, government borrowing gives our governments the ability to act flexibly.

Another look at GDP

If GDP is the amount that a country produces every year, what becomes of the stuff that's produced? It's either used up in the country ('consumption'), retained in the country ('investment') or sent out of the country ('exports'). If you also take into account the amount of imports, there must be a balance.

GDP + Imports = Consumption + Investment + Exports

---

[24] The official figure is $53 billion, but $13 billion is paid by the government to itself as part of the Quantitative Easing programme.

[25] Figure for 2016 estimated based on 2017 debt and interest levels.

So let's see how that breaks down in the case of our three example countries:

Measure	UK	USA	China
GDP	$2.79 trillion	$18.6 trillion	$21.3 trillion
Private consumption	$1.84 trillion	$12.7 trillion	$7.89 trillion
	(66%)	(69%)	(37%)
Government consumption	$0.54 trillion	$3.29 trillion	$2.98 trillion
	(19%)	(18%)	(14%)
Investment	$0.49 trillion	$3.04 trillion	$9.66 trillion
	(18%)	(16%)	(45%)
Exports	$0.80 trillion	$2.23 trillion	$4.68 trillion
	(29%)	(12%)	(22%)
Imports	($0.88 trillion)	($2.73 trillion)	($3.93 trillion)
	(32%)	(15%)	(18.5%)

Yet another look at GDP

One final way of looking at this central measurement of the economy: what sort of activity generates this product?

Measure	UK	USA	China
GDP	$2.79 trillion	$18.6 trillion	$21.3 trillion
Agriculture	$0.02 trillion	$0.2 trillion	$1.8 trillion
	(1%)	(1%)	(9%)
Industry	$0.54 trillion	$3.6 trillion	$8.5 trillion
	(19%)	(19%)	(40%)
Services	$2.23 trillion	$14.8 trillion	$11.0 trillion
	(80%)	(79%)	(51%)

It's a sign of the times that the percentages reported for agriculture are so low. Very few countries now have economies that are primarily based on agriculture: examples are Sierra Leone (71% agricultural) and Somalia (60%). The majority of mature economies are now service economies. Even in China, services have overtaken industry as the chief component of GDP.

Example: defence spending

What use to us are all these numbers? I've tried to present a very small selection of the numbers that might describe, at a very high level, three contrasting economies. You can see, for example, very much lower levels of private consumption in China, and correspondingly higher levels of investment; or the very high dependence on services in the UK and the USA (and increasingly so in the case of China).

I hope that these numbers can also be helpful for you as landmark numbers, to put the numbers in the news into a meaningful context. As an example, let's look at how much each of these countries spends on defence:

Measure	UK	USA	China
GDP	$2.79 trillion	$18.6 trillion	$21.3 trillion
Government spending	$1.10 trillion	$5.66 trillion	$5.50 trillion
Defence spending	$0.051 trillion	$0.611 trillion	$0.404 trillion
	(1.8% of GDP)	(3.3% of GDP)	(1.9% of GDP)
% of Government spending	4.7%	10.8%	7.4%

Example: research spending

For contrast, let's look at how much each country spends on research and development:

Measure	UK	USA	China
GDP	$2.79 trillion	$18.6 trillion	$21.3 trillion
Government spending	$1.10 trillion	$5.66 trillion	$5.5 trillion
R&D spending	$0.045 trillion	$0.47 trillion	$0.41 trillion
	(1.7% of GDP)	(2.6% of GDP)	(2.1% of GDP)
% of government spending	3.6%	7.9%	14.5%

Everybody Counts

Population Growth and Decline

I meet so many that think population growth is a major problem in regard to climate change. But the number of children born per year in the world has stopped growing since 1990. The total number of children below 15 years of age in the world are now relatively stable around 2 billion.   **Hans Rosling**

Which of these is the greatest number?
☐ Population of Chongqing, China
☐ Population of Austria
☐ Estimated world population of blue duiker antelope
☐ Population of Bulgaria

Stand on Zanzibar: How crowded is the world?

There are seven and a half billion people on Earth. Is that a big number?

Stand on Zanzibar is a 1968 science fiction novel by John Brunner. It takes its name from the author's projection that by 2010, if all the world's population stood shoulder to shoulder, the island of Zanzibar could accommodate every one of them. It's a remarkably prophetic book, painting a picture of the then-future that has remarkable resonances with present-day geopolitics and the tone of life today. But how does the premise of its title stand up?

Brunner predicted that there would be seven billion people alive by 2010. In fact, there were 6.9 billion—remarkably close. The calculation that lies behind Brunner's title assumes that each person is allocated a patch of land of 2 feet by

1 foot (in metric terms, about 0.6 metres by 0.3 metres). That's rather small, and would make for a very dense crowd, but let's work with Brunner's numbers for the moment, just to cross-check that they're reasonable. His allocation comes to something less than 0.2 m² per person, which means five people per square metre. That's equivalent to 5 million people per square kilometre.

Zanzibar has an area of 2461 square kilometres, or 2.461 billion square metres. At the density that Brunner specified, Zanzibar could therefore accommodate around 12 billion people, which is well in excess of the seven billion he was correctly forecasting. In fact, 12 billion is a number that exceeds most projections of world population for this century, but we'll come to that.

Now the total land area of our planet is more or less 150 million km², some 60,000 times the area of Zanzibar. If humanity were perfectly evenly distributed over the whole of Earth's land mass, including Antarctica, the Sahara Desert and other remote and inhospitable places, there would be in the region of 50 people for every square kilometre, which means around 20,000 m² for each person. That's just under three soccer fields per person.

People can't live on one-fifth of a square metre, nor do they need three soccer fields each. So how densely do people really live? The port of Macau, a Special Administrative Region in China and sister city to Hong Kong, is the most densely populated area in the world. It accommodates 21,000 people per square kilometre. That works out to about 48 m² per person, which you can visualise as a patch of land measuring 6 m × 8 m. That's not just the allocation of personal space, you understand. It must provide each person with their share of all public space as well: roads, schools, parks,[26] supermarkets, everything. So, if everyone in the world had to live as densely as folk do in Macau, that would mean that the Earth's entire population could fit within just 357,000 square kilometres. Can we **visualise** that?

If we look for a country with approximately that area, we might spot that Japan has a land area of around 365,000 km². So, the land area of Japan would accommodate the whole population of the world, all 7½ billion of us, provided everyone was willing to tolerate the level of crowding that the inhabitants of Macau do.

Or, instead, you could picture a single vast circular city, 675 kilometres across, which would take the better part of a day to drive across at highway speeds. Or, if it's now lodged in your head as a **landmark number**, you might remember that the Earth's equator is 40,000 km long. That lends itself to a

---

[26] Yes, Macau has green spaces too.

visualisation of a kind of ribbon city, girdling the world, 40,000 km in length, but less than 10 km in width. From the midpoint of that ribbon to its edge would be an hour's walk, North or South, and then there would be no habitations at all between you and the Pole.

Picturing it that way, suddenly the world doesn't seem quite so crowded.

But let's face it, this super-city would be extremely congested. Taking as our model the most densely populated place in the world may not be the best idea, and may not even be feasible over a larger area. The logistical problems of supply and removal of waste alone could prove insuperable. How about instead taking an established viable conurbation as the model—in fact, the world's largest, the Pearl River Delta?

This vast built-up area includes both Macau and Hong Kong, is 39,400 km² in size, and is home to a huge number of people. How many people? Estimates are surprisingly imprecise and range from 42 million to 120 million—for our rough calculation, we'll use 80 million. That works out to around 2000 people per km², roughly one-tenth the density of Macau. Each person's share of land is 500 m²: we can visualise that as a plot 25 m × 20 m in size. For them all, we'd need around 3,750,000 km².

So, at a population density comparable to that of the Pearl River Delta, an imagined belt conurbation encircling the world would need to be a little under 100 km wide.

If you prefer to imagine a single circular super-city, it would need to be 2200 km across its diameter. What does that look like? One comparison would be the whole of the European Union (17% too big, with a land area of around 4,400,000 km²). That would accommodate all the world's people at Pearl River density, with room to spare. Or, for a closer fit, taking the countries of Algeria, Niger, and Tunisia together gives us a total land area of more or less 3,800,000 km².

Perhaps the Pearl River Delta is still a little too densely packed for our tastes. What if we allowed ourselves as much elbow room as the average EU citizen now enjoys? The EU is home to around 510 million people, with a total area of 4,400,000 km², giving an average population density of around 120 people per km², one-seventeenth as dense as the Pearl River Delta. At that density, the population of the whole world would need about 65 million km², more or less the area of the seven biggest nations taken together, Russia, Canada, the USA, China, Brazil, Australia, and India.

China itself is slightly more densely populated than Europe, at 145 people per km². The UK has a density a bit more than twice that of the EU's average,

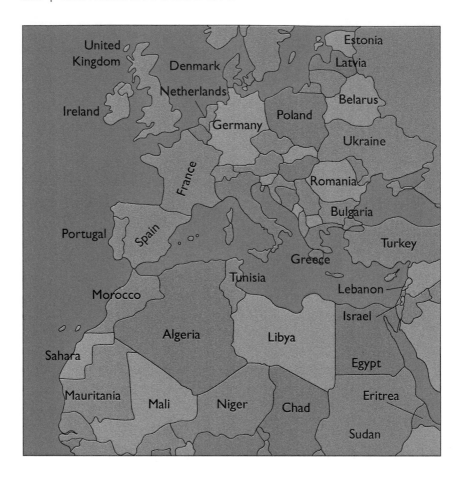

at around 271 people per km². The USA, on the other hand, is relatively sparsely populated, with only 33 people per km². New Zealand is even emptier, with a density of around half that of the USA, around 18 people per km².

So it all depends on your expectations: if you can put up with levels of crowding that are now typical of Europe, there is plenty of room for us all. If you want the open spaces and elbow room that you can find in the USA or New Zealand, then, yes, the world, on average, is more crowded than you'd like.

The rise and rise of *Homo sapiens*

At the dawn of history, around 3000 BCE, when humans invented writing and started to record their story, it's estimated that there were around 45 million people in the world. 3000 years later, at the start of the Common Era, there were around 190 million people. It's now 2017 years after that, and we number 7600 million. There are around 40 people alive now for every person who was alive at the start of the first millennium CE.

But 2000 years is a long time, and that forty-fold increase is equivalent to an average growth rate of a mere 0.18% per annum over those two millennia. Here's a graph showing when that growth occurred:

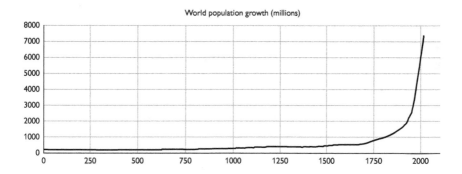

So it's not exactly been a steady process. Let's break that down a little:

• From 1–1000 CE: 0.09% pa.

More or less half the average rate

• From 1000–1700 CE: 0.10% pa.

A slight increase.

• From 1700–1900 CE: 0.50% pa.

A dramatic increase to 5 × the previous rate

• 1900–2000 CE: 1.32% pa.

Another large increase, more than doubling.

You may be reminded, in looking at that graph, of the Moore's law graph we saw in the chapter on logs, and you may be thinking that we need to use a **log scale** to bring some context to these runaway numbers.

Now, as we know, on a log scale, exponential growth should look like a straight line. So, this growth has been beyond exponential: even the power of the log scale fails to tame that dramatic upward movement that's happened in the last few hundred years.

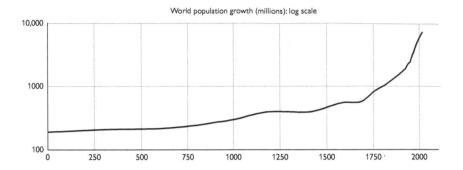

World population growth (millions): log scale

But, looking at recent decades, although the world's population continues to grow, the rate of growth is not as high now as it was through the twentieth century. In fact, the world population growth rate peaked in the 1960s at over 2% pa, but has now declined and is currently around 1.13% pa. If the current rate of growth were to be sustained, the population of the world would be expected to double in around 60 years.

But many demographers think this is unlikely. The reason is that the birth rate in the world has stabilised, and the number of children of age 15 or less has remained more or less constant at around 2 billion since 1990—that's for almost three decades. In Hans Rosling's terms, we have reached 'peak child'. Although we can expect the population of the world to continue to grow, as those generations of children reach maturity and are replaced by roughly equal numbers, the population **growth rate** will continue to decline. Current United Nations predictions project that the world's population will reach a maximum, towards the end of this century, of around 11 billion. That's significantly more than are alive today, but it is nothing like the Malthusian nightmare of exponential growth that seemed to be inevitable back in the 1960s and 1970s.

Landmark numbers
- Peak rate of world population growth (so far): 2%+
- Current rate of world population growth: 1.13%
- UN 2017 mid-range projection of population in 2100: 11.2 billion

---

Human life expectancy

It's not just a high birth rate that drives population growth—it's also the happy fact that, on average, we now live so much longer than we used to. The chart shows how life expectancy has changed over the years in three rather different countries of the world:[27]

Life expectancy, 1543 to 2011
Life expectancy at birth is the average number of years a child born would live if current mortality patterns were to stay the same.

Angola India United Kingdom

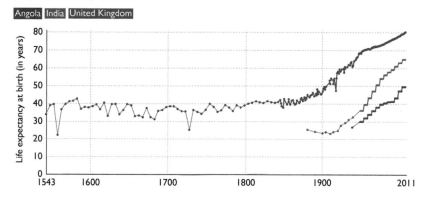

Reproduced under a Creative Commons licence

The dramatic increase in longevity in the twentieth century is very clear to see, and this is not just in the developed Western nations. The chart shows 'Life expectancy at birth'. That's a commonly used measure for length of life, but let's unpack that phrase a little.

- 'At birth' implies that this measure takes account of the rate of infant and child mortality.
- The word 'expectancy' should be treated with caution here—it means something very close to 'average'.

So, the 'life expectancy' is obtained by averaging the predicted future lifespans of a 'cohort' of people born on a given date. For each year, the chart shows a single point to represent that average, but you should imagine these points as marking the centres of much broader distributions, reaching down to zero (to account for infant and child deaths), and up towards older ages (to account for the bulk of people who die at advanced ages).[28]

---

[27] These graphs are from OurWorldInData.org, an inspirational resource for numbers in public life.

[28] As we will see in the next chapter, the average value of a set of numbers is not always the 'typical' value.

Look at the period before 1850. The graph shows that life expectancy hovered around 40 for centuries before then. This does not mean that in those days a 40-year-old person would have been near the end of their life. There would certainly have been people in these communities who were old by our present-day standards. Their long lives would have been balanced by the very short lives of those dying too early. So, the life expectancy averages those who die young with those who die old, and the low average points as much to high rates of child mortality as it does to generally poorer conditions throughout life.

Child mortality

I cannot better these words from the website 'Our World In Data':

One reason why we do not see progress is that we are unaware of how bad the past was. In 1800 the health conditions of our ancestors were such that 43% of the world's newborns died before their 5th birthday…In 1960 child mortality was still 18.5%. Almost every 5th child born in that year died in childhood.

In the form of a graph:

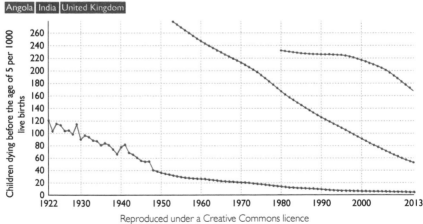

Child mortality, 1922 to 2013
Number of children per 1000 live births who die before reaching the age of 5.

Reproduced under a Creative Commons licence

Just imagine what it must have felt like to live in a society with statistics like those shown for India in the 1950s, where one child in four died before the age of five, compared with the current position in the UK, where the rate is around one in 250. We can call that progress.

It's certainly true that addressing causes of infant and child mortality has done much to increase life expectancy, but it's by no means the only factor. Even setting aside early-years deaths, life expectancy has grown. The chart below shows recent trends and projections in life expectancy among those who have already reached their 10th birthday. In other words, it excludes early childhood mortality.

Life expectancy at age 10, 1950 to 2095
Shown is the number of remaining years a 10-year-old is expected to live. From 2015 onwards, the UN mid-variant projections are shown.

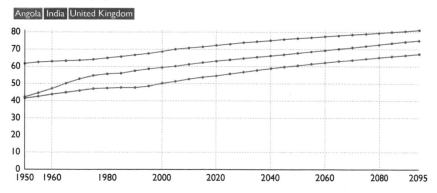

Reproduced under a Creative Commons licence

You can see that in all three countries shown (and they are not exceptional), there is a history and a future projection of a steady rise in life expectancy.

Reductions in child mortality followed by medical advances combatting fatal diseases have resulted in more and more of us living out the full span of our lives. Future medical technology may start extending that span in ways previously only contemplated in science fiction, but the biggest change comes from the fact that more of us reach old age.

We're not the only occupants of this planet

By many measures, *Homo sapiens* is the most successful animal species to have lived on this planet. We have spread to every part of the globe and we outnumber, by far, any animal species of comparable size. 7.5 billion of us are alive today, and roughly 100 billion humans have gone before us.

Collectively, we weigh around 360 billion kilograms—that's more than a quarter of the mass of all land mammals on Earth. That figure is challenged only by domestic cattle at around 500 billion kilograms, with just over 1 billion individuals.

Looking at living creatures on the smaller scale, we're outnumbered. How many ants are there in the world? There's no consensus, even as to the order of magnitude. It's a commonly repeated 'fact' that the biomass of all the ants in the world either exceeds or equals the mass of humans—though this claim does not hold up under scrutiny. When it comes to the number of individuals, though, we can be sure that ants certainly do outnumber us, but we're not even sure by how many orders of magnitude. A recent BBC documentary reported that the estimated number of ants in the world had been revised from 10,000 trillion to a mere 100 trillion, which is still four orders of magnitude (10,000 times) bigger than the human population.

However, the marine population of Antarctic krill may well outweigh us. It's estimated that, in terms of biomass, the krill weigh in at around 500 billion kilograms (roughly the same as all the cattle on Earth).

These numbers, big as they are, seem tiny once we include plant biomass, which brings the total biomass of the Earth to 520 trillion kilograms. That's 400 times the mass of all the Earth's land mammals.

But even that very big number, the mass of everything living on Earth, is just one-fortieth of the mass of the water in North America's Great Lakes. And one-tenth of a billionth of the mass of the whole good Earth. Life on Earth forms a very small part of this planet.

Populations of species

Humans are the most numerous large mammals on Earth,[29] and it should come as no surprise that the species that follow us in that ranking are the animals that we have chosen to domesticate for their muscle power, their meat and milk, their fur and hides, or their companionship. These species are, in descending order of numbers: cattle (1 billion), sheep (1 billion), pigs (1 billion), goats (850 million), cats (600 million), dogs (525 million), water buffalo (170 million), horses (60 million), and donkeys (40 million).

---

[29] Population numbers for small mammals are very hard to establish. Rats and mice may rival humans in number.

Only after these domesticated species do we get to the most numerous wild species, the eastern grey kangaroo at around 16 million individuals,[30] followed by crabeater seals at 11 million. There are in fact many species of seal, several with populations into the millions. It's very hard to count dolphins (and there are 42 species of them), but their numbers also reach into the millions—but remember that's still three orders of magnitude less than humans.

Various species of antelope and buck, including wildebeest and moose, also have numbers in the millions, but these are, relative to the numbers of humans and domesticated animals, very small numbers: a thousand times smaller. Pause to reflect: remember the examples of 'one in a thousand' ratios earlier in the book? Well, there's only one American black bear on this planet for every thousand cows. And one Bactrian camel in the wild for every thousand black bears.

A (descending) number ladder of animal populations

10 billion	Human beings: *Homo sapiens*—7.4 billion
1 billion	Domestic cattle—1 billion
500 million	Domestic dogs—525 million
200 million	Water buffalo—172 million
100 million	Horses—58 million
50 million	Donkeys—40 million
20 million	Eastern grey kangaroos—16 million
10 million	Crabeater seals—11 million
5 million	Blue duiker antelope—7 million
2 million	Impala antelope—2 million
1 million	Siberian roe deer—1 million
500,000	Grey seals—400,000
200,000	Common chimpanzees—300,000
100,000	Western gorillas—95,000
50,000	Bonobo chimpanzees—50,000
20,000	African rhinos (white and black)—25,000
10,000	Red pandas—10,000

[30] Taking four species of kangaroo together, the reds, eastern and western greys, and wallaroos, there were in 2011 around 34 million kangaroos.

5000	Eastern gorillas—5900
2000	Giant pandas in the wild—1800
1000	Wild Bactrian camels—950
500	Ethiopian wolves—500
200	Pygmy hogs—250
100	Sumatran rhinos—100
50	Javan rhinos—around 60

Thanks to the work of many campaigners, there is a general awareness of species that are at risk of extinction. Of course, it is quite proper that conservation efforts should be directed primarily at those species at the bottom of that list, the ones most at risk (and there are dozens of species that could have been chosen to fill those lower slots). But look at how small the population numbers are, even for those species in the middle of the ladder, species that are not on any 'at threat' list, that we don't regard as being at risk.

Take a safari holiday in a wildlife reserve in Southern Africa, and you'll get the impression that there are enormous herds of impala round every corner, in vast numbers. And yet there are only 2 million of them in total—they're outnumbered 4000:1 by humans in the world. The largest cities in the world have ten times that number of humans living in them. We really have squeezed wildlife into a very small portion of the Earth's surface.

We're the ones in charge here

The quicker we humans learn that saving open space and wildlife is critical to our welfare and quality of life, maybe we'll start thinking of doing something about it.

Jim Fowler

In preparing for the writing of this book, nothing has been more sobering than collecting the numbers that quantify the animal populations left on this world. Sadly, by and large, these are not big numbers.

It's not, I think, a question of too many humans: every human life is valuable, and every human alive has the potential to contribute to the great story of humankind. It's a question of too few non-humans. It is shocking to do the comparison and discover that there are 25 times more people living on the tiny Channel Island of Jersey than there are tigers alive in the world. The Madison

Square Gardens concert venue in New York City has two seats for every cheetah left alive. Now that's an interesting **visualisation**!

As we've seen, it's not just the well-known and photogenic endangered species that have numbers that are shockingly low. We thrill to wildlife documentaries showing vast herds of blue wildebeest crossing the African savanna. Impressive until you realise that the city of Nairobi has twice as many people living in it as there are blue wildebeest in the whole world.

Even domesticated animals are not as numerous as you might think. There are just short of 1 billion pigs in the world (half of them in China), despite pigs being such an important food animal. There are half a billion dogs in the world. (Our best friend?—not even one for every 10 people.)

Conservation efforts go into saving those animals that are on the brink of extinction, and that is surely right. But a look at these numbers tells me that that is only the very most urgent of the problems. We are really not doing enough. Mere survival of the species most at risk is a very impoverished ambition. We should have bigger numbers in our sights.

This map, courtesy of the World Bank, shows for each country the proportion of that country that is protected in some way. Darker colours signify a greater proportion that is protected. The world average is 14.8%. The graph

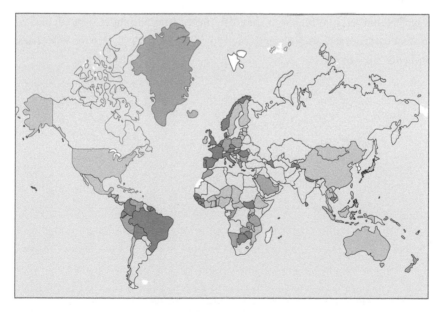

United Nations Environmental Program and the World Conservation Monitoring Centre, as compiled by the World Resources Institute, based on data from national authorities, national legislation and international agreements.

is based on figures for 'terrestrial protected areas' that are 'totally or partially protected areas of at least 1000 hectares that are designated by national authorities as scientific reserves with limited public access, national parks, natural monuments, nature reserves or wildlife sanctuaries, protected landscapes, and areas managed mainly for sustainable use.' That's a mouthful. It also represents a pretty broad definition—too broad for our purposes.[31]

In search of a more valid measure, I turned to a list of the world's largest land-based protected conservation areas. It shows that the largest protected area in the world is in Northeast Greenland, and covers almost a million square kilometres. That's followed by the Ahaggar National Park in Algeria at 450,000 km², and the Kavango–Zambezi Transfrontier Conservation Area, an amalgam of many smaller contiguous parks in the area where the African nations of Zambia, Zimbabwe, Botswana, Namibia, and Angola meet, with 390,000 km². The total of all the protected areas on the list is 4.3 million km². That's only 2.9% of the Earth's land area.

Put another way, if we humans could release just 3% more of the Earth's land area back to the wild, we would at a stroke more than double the area available to the wild animals of the world. Of course, I oversimplify. Such a change would be very disruptive to human populations and would pose tremendous political challenges. It could never be achieved in a straightforward way.[32] Nonetheless, it seems an absurdity, and extremely short-sighted, that we protect and preserve such a small fraction of the planet's surface for wilderness and wild creatures.

[31] Indeed, it includes the area where I live. It's an area called the 'Surrey Hills' in England and it's rural farming country, but it's very far from being a wildlife preserve.

[32] Perhaps it's not as overambitious as all that. Some go even further in their proposals, such as the biologist E. O. Wilson, who is campaigning for half of the Earth to be set aside for wildlife.

Power to the People?

Population of **Austria** (8.71 million people) is
   about as big as the population of **Lima, Peru** (8.69 million people)

Population of **Pakistan** (202 million people) is
   2.5 × population of **Germany** (80.7 million people)

Population of **Puerto Rico** (3.58 million people) is
   4 × estimated world population of **African buffalo** (890,000)

Population of **Barbados** (291,500 people) is
   50 × estimate of number of **Eastern gorillas** (5880)

Estimated population of **African elephants** (700,000) is
   100 × estimated population of **Arabian oryx antelope** (7000)

Population of **Jakarta, Indonesia** (10.07 million people) is
   1000 × estimated world population of **Sunda clouded leopards** (10,000)

Population of **Western Sahara** (587,000 people) is
   about as big as population of **Luxembourg** (582,000 people)

Population of **Morocco** (33.7 million people) is
   100 × population of **Iceland** (336,000 people)

Measuring How We Live

Inequality and Quality of Life

Which of these countries scores highest for happiness?
- ☐ United Kingdom
- ☐ Canada
- ☐ Costa Rica
- ☐ Iceland

Measuring variation and inequality

Welcome to Lake Wobegon, where all the women are strong, all the men are good-looking, and all the children are above average.   **Garrison Keillor**

— In the UK in 2015, average after-tax income was around £24,000. *Is that a big number?*

— In Canada, the poorer 50% of the population own 12% of the wealth. *Is that a big number?*

— In 2015, 12% of the world's population were living in extreme poverty. *Is that a big number?*

The Stealth roller coaster at Britain's Thorpe Park is a based on a very simple concept. The riders are launched forward and almost immediately start to ascend a vertical track. Without further propulsion, they go over the top (62 metres high) and then immediately are taken into a vertical descent. It's rather scary.[33] The

---

[33] Believe me, it is—I've taken that ride. Why it's called Stealth is anyone's guess!

train makes another shallow rise and then slows to complete the circuit. The whole ride takes a mere half minute and the track is half a kilometre long—that makes the average speed equal to 60 km/h, which is hardly impressive in itself. But the real point is the variation in speed. The rider is accelerated from a standing start to 129 km/h in under two seconds, loses much of that speed on the ascent, and then regains it in the terrifying drop. The changes in speed are felt as G-forces, and that's part of what makes it thrilling. Averages don't always tell the full story.

In Houston, Texas, average wind speeds for August are around 5.5 km/h. But on 26 August 2017, Hurricane Harvey made landfall with wind speeds in excess of 200 km/h. When preparing for disasters, averages aren't much use. You need to know how weather conditions vary, and hope that you get the extremes right.

And that's the case with many sets of numbers: sometimes the average is meaningless, and sometimes it's downright misleading. The average after-tax income of a taxpayer in the UK in 2013 was just over £24,000. But most earners (in fact, 65% of all earners) had an after-tax income that is much less than this average. And the reason for this is because when it comes to incomes, we're dealing with a **skewed distribution**.

Here's a bar graph of that incomes dataset, showing what percentage of the population falls into each of nine income bands (UK taxpayer income after tax in 2013):

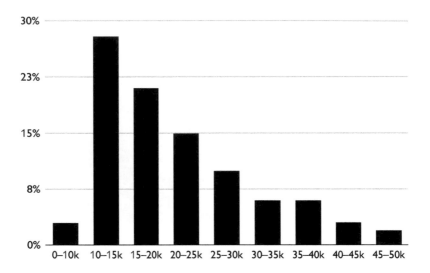

The skewness is clear: there are plenty of taxpayers on low incomes, a few who are on larger incomes, and a very few (not shown—they're too small a number to graph effectively) with very large incomes indeed. Inevitably and mathematically, the effect of those few very large cases is to drag the average up. It's plain to see that while £24,000 may be the calculated average, it's certainly not a typical figure.[34]

We use averages in part because they are so easy to work with. Averages make rule-of-thumb calculations simple, and in researching and writing this book, I have shamelessly used averages in all sorts of places to simplify big numbers and bring them down to human scale. But averages have a homogenising effect, with the implicit assumption that whatever base (such as the tax-paying population) is used to calculate the average, the members of that base are all the same. Whenever we do a per capita calculation, something that is part of our toolkit as numerate citizens, we are calculating averages, and so we run the risk of missing the variability in the data that we are looking at. Very

---

[34] Another way of demonstrating the skewness of the data: if we were to leave off the top 10% and the bottom 10% of that distribution of taxpayers (which in a symmetrical distribution should leave the average unchanged), the average income drops to £21,000. This shows that the top 10% are 'pulling upward' much more strongly than the bottom 10% are 'pulling downward'.

often that is the best that we can do, and usually it is a worthwhile first step, but we should be aware of the perils.

Every statistician's toolbox includes a set of standard descriptive statistics to summarise data sets. First among these is the **mean**, which is what most of us would call the 'average'. Next is the **variance**, which, as its name suggests, is a measure of how much the data varies above and below the mean. The third is the **skewness** of the distribution, a measure of lopsidedness.[35] For a statistician, these descriptive statistics can provide a great deal of insight, but for a non-specialist, these calculated numbers can be difficult to interpret. If we use the mean, we lose a lot of information. Even worse, we may end up giving false impressions.

When reporting on average taxpayer incomes, what we usually want to convey is some sense of what is **typical**. For this purpose, the mean is misleading. A preferred statistic in this case would be the **median** income, which is the number that splits the data set in two: as many cases fall below the median as fall above it. This approach can be generalised, by looking at how many of the cases fall into different percentile bands, that is, how much do the lowest 10% of taxpayers earn, how much do the next 10% earn, and so on.

Going back to the data for UK incomes, looking at all the data, the median taxpayer income is £19,500. If we line up all the members of a group in order, the one at the midpoint is the median. 50% of cases are below this value: it's known as the 50th percentile. To understand more about the variability in data, you can look at other percentiles. In the case of the 2013 taxpayer incomes, we can show the following:

Percentile	Income	Note
10th	11,400	10% of taxpayers had incomes less than this
20th	13,100	10% had incomes between this and 11,400
30th	14,900	
40th	17,000	
50th	**19,500**	The median (half earned more, half less)
60th	22,600	

[35] These are called the moments of the distribution. And it doesn't stop there: **kurtosis** is the fourth in this series and measures how heavy the tails of the distribution are. Beyond that, interpretations of the meaning of the moments become ever more difficult.

Percentile	Income	Note
64th	24,000	The mean (the 'average')
70th	26,600	
80th	32,600	80% earned less than this
90th	41,500	10% earned more than this
99th	107,000	'The 1%' earned more than this

It's clearly not possible to do an analysis like this for every statistic that we see in the news, but try to remember, next time a politician quotes 'a rise in average incomes', that what he is talking about may not affect all taxpayers equally.

Is that a big country?

The United Nations has 193 member nations. Five of these are permanent members of the Security Council, and, by virtue of this status, have superior

powers, but all the other nations are treated equally and have equal votes in the General Assembly. When the competitors march in their national teams at the opening ceremony of the Olympics, they get equal treatment—the same size banners, the same public announcements. Fairness seems to argue that each country should get the same treatment. But when it comes to actual economic and political clout, it's clear that not all countries carry the same weight. We know there are big countries and small countries, but perhaps we don't fully understand quite how skewed that distribution is.

Consider the chart below, showing how many of the countries in the world have population counts that fall into the bands shown. More than half the countries in the world have populations that are less than 10 million, only 13 have populations above 100 million.

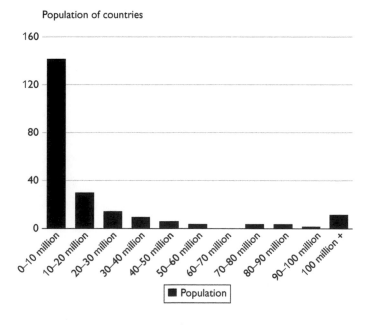

The average population of all the countries in the world is around 32 million people. But that number can hardly be called the midpoint of the chart shown above. Only 40 countries have larger populations than that, and 189 countries have populations smaller than 32 million. Once again, for an unbalanced distribution, the average is not a useful measure. In this case, the median figure is around 5½ million. Finland and Slovakia are close to this 'typical country' size.

Measuring inequality—the Gini index

In 1912 Corrado Gini, a sociologist and statistician, introduced a measure for quantifying 'statistical dispersion'—or variability—in a set of data. The Gini Index is quite simple to calculate, and, with a single number, captures the 'unfairness' of a distribution. It's most often used to measure inequality of income and wealth distribution. A Gini value of zero means complete equality—every person has an equal share. A value of 100 means complete inequality—one person has everything.

Here's a selection of Gini values for income inequality across a selection of different countries, ranging from the most equal to the most unequal. This table also shows a correlated expression of inequality: the percentage of national income that goes to the bottom 50% of the population. For a country with no inequality at all, this measure would be 50%.

Country	Gini Index	Bottom 50% receive x% of the income
Denmark	24.8	Bottom 50% receive 34.2% of the income
Sweden	24.9	Bottom 50% receive 34.1% of the income
Germany	27.0	Bottom 50% receive 32.9% of the income
Australia	30.3	Bottom 50% receive 31.0% of the income
Canada	32.1	Bottom 50% receive 30.0% of the income
UK	32.4	Bottom 50% receive 29.8% of the income
Poland	34.1	Bottom 50% receive 28.9% of the income
Japan	37.9	Bottom 50% receive 26.8% of the income
World	38.0	Bottom 50% receive 26.8% of the income
Thailand	39.4	Bottom 50% receive 26.0% of the income
Nigeria	43.7	Bottom 50% receive 23.8% of the income
USA	45.0	Bottom 50% receive 23.1% of the income
China	42.2	Bottom 50% receive 24.6% of the income
Brazil	51.9	Bottom 50% receive 19.7% of the income
South Africa	62.5	Bottom 50% receive 14.8% of the income

The figures shown above relate to inequality of **income** distribution. Inequality of **wealth** is a different matter. The disparity is much greater. Here is the same list, but sorted into order of wealth inequality:

Country	Gini Index	Bottom 50% of have x% of the wealth
Japan	54.7	Bottom 50% have 18.4% of the wealth
China	55.0	Bottom 50% have 18.2% of the wealth
Brazil	62.0	Bottom 50% have 15.0% of the wealth
Australia	62.2	Bottom 50% have 14.9% of the wealth
Poland	65.7	Bottom 50% have 13.4% of the wealth
Germany	66.7	Bottom 50% have 12.9% of the wealth
Canada	68.8	Bottom 50% have 12.0% of the wealth
UK	69.7	Bottom 50% have 11.7% of the wealth
Thailand	71.0	Bottom 50% have 11.1% of the wealth
Nigeria	73.6	Bottom 50% have 10.0% of the wealth
Sweden	74.2	Bottom 50% have 9.8% of the wealth
South Africa	76.3	Bottom 50% have 8.9% of the wealth
USA	80.1	Bottom 50% have 7.4% of the wealth
World	80.4[36]	Bottom 50% have 7.3% of the wealth
Denmark	80.8	Bottom 50% have 7.1% of the wealth

There's a lot to think about in these numbers. To begin with, note that wealth inequality as shown in this table is far greater than the income inequality in the previous table. And note these curiosities: Denmark is the most equal country in terms of income, and yet very unequal in terms of wealth.[37] Japan is close to the world average for income inequality, but is the most equal country of all when it comes to wealth.

Quality of life: Millennium Development Goals

What gets measured, gets managed.   Attributed to Peter Drucker

In 2000 CE, the United Nations made a Millennium Declaration, committing to eight international development objectives. 189 world leaders signed up to

[36] Note that this is not an average of the Gini coefficients of the countries—it's the overall global inequality that's being measured. Inequality between countries is what makes this so large.
[37] Only Namibia and Zimbabwe have greater inequality of wealth distribution.

the declaration. What makes this set of goals interesting from a numbers point of view is that eight specific goals emerged from this declaration, and that these had specific quantified sub-targets and a specific end date, namely 2015. In other words, they were testable.

Predictably, these goals did not meet with universal agreement, and were the outcome of political compromise. At the time, and subsequently, they were much criticised. But, by and large, the criticisms were to do with the selection of goals—many groups were disappointed that their particular interests were not included—or with the relative priorities given to the goals. Few argued that the targets in themselves were unworthy.

Specific targets, a definite deadline: what more could a numerate enquiring mind hope for? Let's have a look at how things turned out.

Goal 1: Eradicate extreme poverty and hunger

Target 1.A: Halve, between 1990 and 2015, the proportion of people living on less than $1.25[38] a day

Achievement: Overachieved

In 1990, nearly half of the population in the developing world lived on less than $1.25 a day. That proportion dropped to 14% in 2015. In the world as a whole, the proportion dropped from 36% to 12%.

Target 1.B: Achieve decent employment for women, men, and young people

Achievement: Very poor

In developing regions, over the period 1991–2015, the percentage of the population in employment **fell from 64% to 61%**. In developed regions, it fell from **57% to 56%**. Still, the number of people in the working middle class—living on more than $4 a day—has almost trebled between 1991 and 2015. This group now makes up half the workforce in the developing regions, up from just 18% in 1991.

Target 1.C: Halve, between 1990 and 2015, the proportion of people who suffer from hunger

---

[38] Adjusted for inflation. This is equivalent to a dollar a day in 1996 money.

Achievement: Very good progress

The proportion of undernourished people in the developing regions has fallen by almost half since 1990, from 23.3% in 1990–1992 to 12.9% in 2014–2016.

> **Landmark number** In 2015, 12% of people were living in extreme poverty, down from 36% in 1990.

Goal 2: Achieve universal primary education

Target 2.A: Ensure that, by 2015, children everywhere, boys and girls alike, will be able to complete a full course of primary schooling

Achievement: Good progress

The primary school net enrolment rate in the developing regions reached an estimated 91% in 2015, up from 83% in 2000. (Of course, the goal of 100% is in practice unachievable.) The figure for developed regions held steady at 96%.

> **Landmark number** In 2015, 91% of children in developing regions were enrolling in primary education.

Goal 3: Promote gender equality and empower women

Target 3.A: Eliminate gender disparity in primary and secondary education, preferably by 2005, and in all levels of education no later than 2015.

Achievement: Excellent progress

About two-thirds of countries in the developing regions have achieved gender parity in primary education. Even where they have not achieved parity, they have moved much closer to it. Looking at the gender split of workers in wage employment in the non-agricultural sector, this has moved from women forming 35% of the workforce in 1990 to 41% in 2015.

Goal 4: Reduce child mortality

Target 4.A: Reduce by two-thirds, between 1990 and 2015, the under-five mortality rate

Achievement: Good progress

The global under-five mortality rate has declined by more than half (53%), dropping from 90 to 43 deaths per thousand live births between 1990 and 2015.

> **Landmark number** In 2015, child mortality was 4.3%, down from 9% in 1990.

Goal 5: Improve maternal health

Target 5.A: Reduce by three-quarters, between 1990 and 2015, the maternal mortality rate

Achievement: Moderate progress

Since 1990, the maternal mortality rate has been cut nearly in half (45% reduction).

Target 5.B: Achieve, by 2015, universal access to reproductive health

Achievement: Slow progress

Only half of pregnant women in developing regions receive the recommended minimum of four antenatal-care visits. Worldwide, the proportion of women aged 15–49, married or in a union, who were using any method of contraception increased from 55% in 1990 to 64% in 2015.

Goal 6: Combat HIV/AIDS, malaria, and other diseases

Target 6.A: Have halted by 2015 and begun to reverse the spread of HIV/AIDS

Achievement: Moderate progress

New HIV infections fell by approximately 40% between 2000 and 2013, from an estimated 3.5 million cases to 2.1 million. 2.4 million deaths from AIDS were recorded in 2005, the peak year. By 2013, this had reduced to 1.5 million deaths. The number of children orphaned by AIDS peaked in 2009.

Target 6.B: Achieve, by 2010, universal access to treatment for HIV/AIDS for all those who need it

Achievement: Good progress

By June 2014, 13.6 million people living with HIV were receiving antiretroviral therapy (ART) globally, an immense increase from just 800,000 in 2003. ART averted 7.6 million deaths from AIDS between 1995 and 2013.

Target 6.C: Have halted by 2015 and begun to reverse the incidence of malaria and other major diseases

Achievement: Excellent progress

Between 2000 and 2015, the global malaria incidence rate fell by an estimated 37%, and the global malaria mortality rate decreased by 58%. Between 1990 and 2013, the mortality rate due to tuberculosis fell by 45%.

Landmark number Annual AIDS deaths are believed to have peaked in 2005 at 2.4 million.

Goal 7: Ensure environmental sustainability

Target 7.A: Integrate the principles of sustainable development into country policies and programmes; reverse loss of environmental resources

Achievement: Mixed, mostly poor

Deforestation has slowed to around 60% of the rate it was in the 1990s. Greenhouse gas emissions continue to increase, and are now 50% higher than the 1990 level. The ozone layer is expected to recover by mid-century, thanks to concerted global efforts to eliminate ozone-depleting substances. Overexploitation of marine fisheries is rising. Water scarcity affects more than 40% of the global population and is projected to rise.

Target 7.B: Reduce biodiversity loss, achieving, by 2010, a significant reduction in the rate of loss

Achievement: Mixed, mostly poor

The Red List index, which measures trends in species' risk of extinction, shows that a substantial proportion of species in all taxonomic groups examined to date are declining overall in population and distribution. This means they are increasingly threatened with extinction. Global coverage of protected areas has expanded since 1990, and protected areas are projected to reach at least 17% of terrestrial areas and inland waters and 10% of marine and coastal areas by 2020.

Target 7.C: Halve, by 2015, the proportion of the population without sustainable access to safe drinking water and basic sanitation

Achievement: Mixed—excellent for drinking water, moderate for sanitation

Between 1990 and 2015, the proportion of the global population using an improved drinking water source has increased from 76% to 91%, surpassing the target, which was met in 2010. The proportion of people without access to improved sanitation fell from 46% to 32%.

Target 7.D: By 2020, to have achieved a significant improvement in the lives of at least 100 million slum-dwellers

Achievement: Steady progress

The proportion of the urban population living in slums in the developing regions fell from approximately 39.4% in 2000 to 29.7% in 2014.

Goal 8: Develop a global partnership for development

Target 8.A: Develop further an open, rule-based, predictable, non-discriminatory trading and financial system

Achievement: Moderate progress

The proportion of developed country imports originating from developing countries that are admitted duty-free has significantly increased over the last 15 years (from 54% to 79%).

Target 8.B: Address the special needs of the least developed countries (LDCs)

Achievement: Poor, very little positive change

Target 8.C: Address the special needs of landlocked developing countries and small island developing states

Achievement: Poor, very little positive change

Target 8.D: Deal comprehensively with the debt problems of developing countries through national and international measures in order to make debt sustainable in the long term

Achievement: Initially encouraging, but predicted to deteriorate

In 2013, the debt burden of developing countries was 3.1%, measured as a proportion of external debt service to export revenue. This was a major improvement over the 2000 figure of 12.0%.

Target 8.E: In cooperation with pharmaceutical companies, provide access to affordable, essential drugs in developing countries

Achievement: No real information available

Target 8.F: In cooperation with the private sector, make available the benefits of new technologies, especially information and communications

Achievement: Dramatic progress, but increasing inequality

As of 2015, 95% of the world's population is covered by a mobile–cellular signal. Internet penetration has grown from just over 6% of the world's population in 2000 to 43% in 2015, covering 3.2 billion people. While for developed countries this figure is around 82% of the population, for developing nations it is around one-third of the population. In Sub-Saharan Africa, the figure is around 20%.

How does the overall achievement measure up?

This is a report card with mixed results. The goals themselves are a mixed bag— it's clear from the outcomes that they've not all turned out to be easily measurable. In some cases, they seem to be wishes more than goals.

On the other hand, some of the achievements are remarkable. In particular, the progress towards some of the most fundamental goals is outstanding: the reduction in poverty and hunger, the improvement in child mortality and provision of primary education, and the treatment of disease. Fewer people dying, fewer living in poverty and hunger, the prevalence of disease being rolled back: this feels like things getting better.

Quality of life: Human Development Index

Danes pay very high taxes, but in return enjoy a quality of life that many Americans would find hard to believe.   Bernie Sanders

In 1990, the economist Mahbub ul Haq, working with Nobel Prize winner Amartya Sen, devised and launched the Human Development Index (HDI), a United Nations–backed measure of, well, human development. The aim was to refocus measures of development in order to concentrate on people-centred policies.

The HDI combines three components: life expectancy, education, and earnings. In other words, it is a single-figure answer to the question 'Does the average person in this or that country enjoy long life, good education, and good income?'

As currently defined, a perfect score would be achieved by a country that boasted a life expectancy of 85 years, an average of 15 years of schooling in the population, an expectation of 18 years of education for those starting school, and a per capita income of $75,000 a year.

The HDI is based on averages, and averages can be misleading for skewed data sets. For this reason, a variation of the HDI, the Inequality-Adjusted HDI (IHDI), was introduced in 2010 to incorporate the effects of inequality in each country. The HDI can be thought of as measuring the **potential** for a good life in each country, while the IHDI tries to measure the country's **achievement**, by 'penalising' inequality in health, education, and income.

Norway reliably tops the HDI (and IHDI) chart. Its HDI for 2015 was expected to be 0.949. At the bottom of both tables is the Central African Republic. Here's how other countries compare with these:

Country	HDI	IHDI
Norway	0.949	0.898
Australia	0.939	0.861
Denmark	0.925	0.858
Germany	0.926	0.859
USA	0.920	0.796
Canada	0.920	0.839
Sweden	0.913	0.851
UK	0.909	0.836
Japan	0.903	0.791

Country	HDI	IHDI
Poland	0.855	0.774
Brazil	0.754	0.561
China	0.738	0.543[39]
Thailand	0.740	0.586
World	0.717	N/A
South Africa	0.666	0.435
India	0.624	0.454
Nigeria	0.527	0.328
Central African Republic	0.352	0.199

While these measures can be criticised, the HDI and IHDI nonetheless represent a serious attempt to measure at least some of the things that most would agree go towards making 'a good life'.

In the 2015 edition of this index, it is remarkable that there are only 13 out of 188 countries where the HDI went down from the previous year's value. In a further 11 countries, the index remained constant. The remaining 164 countries all logged an increase in the index.

Can we use the word 'progress'? It's not a fashionable term these days, but (with some reservations) this feels like progress to me. Through the ages, people have harked back to 'the old days' as somehow being better than the present. There's always a golden age lurking in the vague memory. These figures serve as a counter to that view. For all that's still wrong in the world, I feel that these numbers are cause for optimism.

Quality of life: Happiness Index

It's a concept that brings you up short in its simplicity: to assess quality of life, just ask people how happy they are. This may seem childishly naive, but then we know that happiness does not correlate perfectly with other, more objective, measures of life circumstances.

---

[39] 2012 data.

So, in 2011, the United Nations initiated a process of information gathering that led in 2012 to the creation of the first World Happiness Report. Happiness scores were assessed around the world, as were six candidate explanatory factors:

- GDP per capita
- Social support
- Healthy life expectancy
- Freedom to make life choices
- Generosity
- Trust

Additionally, there was a 'residual' or unexplained factor that reflected the amount of happiness score that could not be accounted for by the six explanatory factors—an X-factor that contributed to national happiness.

Here's how our sample of countries rated:

Country	Happiness Index	Unexplained
Norway	7.54	2.28
Denmark	7.52	2.31
Canada	7.32	2.19
Australia	7.28	2.07
Sweden	7.28	2.10
USA	6.99	2.22
Germany	6.95	2.02
UK	6.71	1.70
Brazil	6.64	**2.77**
Thailand	6.42	2.04
Poland	5.97	1.80
Japan	5.92	**1.36**
China	5.27	1.77
Nigeria	5.07	**2.37**
South Africa	4.83	**1.51**
India	4.32	1.52
Central African Republic	2.69	2.07

Once again, Norway comes top of the league. They must be doing something right. And once again, the Central African Republic is at the bottom of the table. The Unexplained column is intriguing. South Africa and Japan are the places where people are gloomier than might be expected (at least when looking at the six explanatory factors). And Brazil and Nigeria have the top scores in the Unexplained column—places where people are happier than they have a right to be.

The Standards of Life

GDP per capita of **Macau** ($110,000) is
   about the same as the GDP per capita of the **United Arab Emirates** ($109,000)

GDP per capita of **New Zealand** ($37,500) is
   about the same as the GDP per capita of **Greenland** ($37,600)

GDP per capita of the **Czech Republic** ($31,750) is
   20 × the GDP per capita of **Rwanda** ($1575)

GDP per capita of **Norway** ($67,800) is
   10 × the GDP per capita of **Bolivia** ($6800)

The **Gini Index for Income (Inequality)** in **Bangladesh** (32.1) is
   exactly the same as the **Gini Index for Income** in **Canada** (32.1)

The **Human Development Index** in **Ireland** (0.923) is
   almost the same as the **Human Development Index** in **Iceland** (0.921)

The **Happiness Index** in **Australia** (7.284) is
   exactly the same as the **Happiness Index** in **Sweden** (7.284)

Summing Up

Numbers Still Count

My big thesis is that although the world looks messy and chaotic, if you translate it into the world of numbers and shapes, patterns emerge and you start to understand why things are the way they are.   **Marcus du Sautoy**

Which of these is most true?
- ☐ The end is nigh
- ☐ It was better in the old days
- ☐ We'll muddle through
- ☐ We're heading for the sunlit uplands

There's no answer in the back of the book for this quiz.

Numbers are natural

Numbers and our sense of number arise naturally from the world and from life. As the first humans began to understand, describe, and control their lives and the world, they started using numbers. Numbers proved essential for organising societies and for building our astonishingly complex, confusing, fascinating world. Numbers were the keys that unlocked our understanding of nature and allowed us to harness the potential of physics, chemistry, biology, and other sciences. Numbers allowed *Homo sapiens*, a modest ape species, to become an overwhelming presence on this planet, and more, to form a single global community, with a collective body of knowledge and understanding that may be unique in the universe.

But lately the numbers seem to be escaping our grasp. There is more information available to us than we could ever comprehend. Big data fuels number-crunching algorithms that invisibly shape our lives. Should we be worried? Why do we still need to bother about numbers? Does numeracy still matter?

Can't computers handle all this numbers stuff?

For decades, computers have been handling routine bookkeeping, and yet we still employ accountants. For decades, computers have done the hard numbers work for engineers, for architects, for actuaries, for all numerate professions. The slide rule, that indispensable tool of engineers just 50 years ago, is now archaic. Yet those professions have not disappeared. The computer is still a tool, but as that tool becomes more and more powerful, it starts acting more like a partner.

Where will you find the best player of chess in the world today? And no, it's not a computer—well, not only a computer. True, IBM's Deep Blue beat Garry Kasparov in 1997. After that defeat, though, Kasparov gave some deep thought to the implications of his loss. He reflected that that humans and computers tackle the game of chess in very different ways, and from this insight Kasparov came up with the idea of Advanced Chess, also known as Freestyle Chess, which is played between human–computer 'teams'. These teams, sometimes involving multiple humans and multiple computers, are now the world's best 'players' of chess.

This gives a hint at a possible accommodation between man and machine: cooperation rather than competition. And, in this collaboration, one of the roles that humans will play will be to use our number sense to connect the abstract numbers to practical reality.

Haven't the number-crunching experts let us down?

Infamously, in the pre-Brexit discussions, Michael Gove (at the time a UK government minister) argued: 'We've had enough of experts'. That statement should bring you up short. It's outrageous for a senior minister to so lightly dismiss those with the most informed opinions available. He would surely not reject expertise in his plumber, or in his surgeon.

But I can understand why his argument resonates. Numerate experts have scarcely been infallible in the past. There are plenty of fields where experts disagree publicly, perhaps most visibly in regard to economics and medicine. But science advances through disagreement, even through failure. That is its strength; it is a self-healing system. And so, steadily, inexorably, we evolve towards a better, more accurate understanding of where truth lies. To dismiss expert knowledge is to misunderstand expertise. It is ignorant, and irresponsible, and betrays a lack of comprehension of how knowledge is acquired.

But the experts can't even make up their minds!

Errors using inadequate data are much less than those using no data at all.

Charles Babbage

Don't reject numbers for lack of precision or certainty. An imprecise number is not a worthless number. Experts in subjects like economics can never make precise predictions—after all, their subject is human behaviour. But even if there is some fuzziness in the numbers, this will nonetheless often represent a sound basis for robust decision-making. Knowing that the outcome of some policy is uncertain, and perhaps even being able to measure that uncertainty, may be essential to making the right choice. Part of being numerate is being able to engage critically with uncertain material, to make better judgements about where trust is warranted, and to know when extra investigation or extra scrutiny is called for.

By all means be sceptical of the experts. But don't dismiss them out of hand—look to apply the test of **cross-comparison**. Ask if the numbers being presented are reasonable. Through numeracy, put yourself in a position to understand where their arguments are weak, and where they are strong. You can develop your own expertise: certainly enough expertise to detect snake oil. 'But you can't be sure of that' should never be a strong enough argument to dismiss an expert opinion.

But you can prove anything with numbers, can't you?

It is easy to lie with statistics; it is easier to lie without them.   **Frederick Mosteller**

Not quite: it's true that bad arguments can be bolstered by numbers, but this is really an argument for greater numeracy. Real-world numbers form a coherent network, and connect with reality in many places. Trace through the network until you find a number that you can validate independently, and it's possible you'll find a discrepancy. The more numerate we are, the better chance we have of spotting deceit and error.

Five techniques

How do we uncover a number-based deceit? By doing cross-comparisons, checking for plausibility and looking for inconsistencies. By asking: 'Is That A Big Number, given what I already know?' And by now you have the techniques for doing this:

- A few well-chosen **landmark numbers** will give you the measuring sticks you need to establish a **context**.
- **Visualisation** will help you form a view of whether or not the numbers seem **reasonable** or not.
- By **dividing and conquering**, you will cut through complex situations and reduce the questions to **simpler forms**.
- Using **rates and ratios** will reduce big numbers to **human-scale** equivalent numbers within your comfort zone.
- **Log scales** will let you make meaningful comparisons between numbers that are **very different in scale**.

Knowledge is power

It's easy to lie with statistics. It's hard to tell the truth without statistics.

Andrejs Dunkels

Through the Internet, we now have access to more information than our parents could ever have imagined. This book could not have been written without the

Internet: there is very little here that could not be found with a few minutes of Googling. Perhaps there is no longer a need to know facts, because the Internet is now an ever-present resource: an extension of our own brains. Wikipedia will do the remembering for us. No?

No. We're truly not there yet, and I suggest we never will be. To **know** something is to have it instantly available in your mind, not just as a bare fact, but embedded in a context of meaning and associations. Even if your knowledge is imperfect and imprecise, and you have no more than a vague sense of the scale of the important numbers, that will often be enough to empower you to call out deception. And the human brain is as yet unmatched when it comes to making connections between disparate items of knowledge, and synthesising something new from them.

Life is becoming more complex: the numbers are becoming bigger, and there are more of them. One possible response to this is to allow ourselves to become comfortably numb. If we do that, we resign ourselves to powerlessness. Truth becomes a devalued currency. We will drift on the current, prey to any plausible huckster.

Or we can choose to engage, and seek to understand, and use numbers to gain a sense of confidence in what we know, and provide an underpinning for our beliefs and values.

Life is chaotic. But life is good

None of us know all the potentialities that slumber in the spirit of the population, or all the ways in which that population can surprise us when there is the right interplay of events.   Vaclav Havel

In the introduction to this book, I described a recurring image of a body of water. The surface moves chaotically, and it's hard to see which way the bulk of the water is moving. I likened this to trying to understand whether things are getting better or getting worse in the world.

In fact, I'm in no doubt as to which way the water is moving. With a couple of exceptions, it's flowing the right way. We may have bad years, but we've had and are having good decades. People are living longer. People are living healthier lives. People receive better education. The Millennium Development Goals may have their justified critics, but for all their imperfections they demonstrate that things are changing, and many things are changing for the better. The same trends are clear when you look at the Human Development Index.

If I could pick a time in history to live my life, would I rather be 20 years old in 1817, 1917, 1967, or 2017? Undoubtedly, now. Exciting though the 1960s were, and although the freedoms brought by the social and technological advances over the past decades have made the last 50 years an extraordinarily exciting time to live through, the possibilities that lie ahead are even greater. The choices that technology can open up for human creativity and achievement are literally unimaginable: the next 50 years will be a fantastic ride. Scary, yes, but exciting.

What lies behind the improvements we have seen in recent years in the condition of humanity? In a word, science. Abatement of disease, improvements in food production and preservation, communication and education. The word 'progress' is out of favour. We've been sold the line that science causes more problems than it solves, but that's not true. In fact, the quality of life has been raised higher than ever before for more people than ever before.

But even if the rising tide of progress has floated everyone a little higher, the benefits have not been equally shared. This limits the potential of billions. It's wasteful of creativity, of human energy, and of lives. To our shame, many people still live in inhuman conditions. But there are fewer than there were last year (and there will be fewer still next year). And if we feel shame, as we should, for the unequal sharing of what the world offers, and for the destruction we've caused to the natural world, and for the wars that still rage, yet we are also entitled to feel extraordinary pride for the billions who are now able to live healthy, meaningful lives of unquestionable value.

We have had an irreversible damaging impact on our environment: the climate, the resources, the animal populations of our planet. No other issue competes with this one in terms of potential long-term consequences and costs. We won't escape from this unscathed, or without leaving behind an awful trail of damage. But survive? Yes, we will survive. And who will bring about that survival? Those with a clear and coherent worldview that connects to reality. Those who can see how the deep waters flow and are able to ignore the froth and foam floating on the top. The numerate citizens of this world.

The human story is full of horrors and wonders. We've done wicked deeds and we've had glorious achievements. But, above all, we progress. We build on what has gone before. We write the story of humanity. We don't always learn as much as we could from our past mistakes, but, by and large, things are getting better. As a whole, the positive contributions of billions of individual lives have outweighed the negatives.

This is not a smooth and unbroken march into a uniformly positive, predictable, planned-out future. Life remains chaotic and confusing. But every one of the 7+ billion humans alive today has the potential to contribute. Every child that is born carries hope for the future. Every extra year of life lived in good health is a chance to write a chapter in the human story.

Life is good. Make it count.

Aren't you glad to know that...?

The length of the **River Thames** (386 km) is
    2 × the length of the **Suez Canal** (193.3 km)

The age of **earliest Neanderthal fossils** (350,000 y) is
    10 × the age of the **earliest cave paintings** (35,000 y)

The estimated world population of **hooded seals** (662,000) is
    250 × the population of **Svalbard** (2,640)

The diameter of **Saturn's rings** (282,000 km) is
    twice (2 ×) the diameter of **Jupiter** (140,000 km)

The mass of a **Gulfstream G650 business jet** (45,400 kg) is
    100 × the mass of a **concert grand piano** (450 kg)

The overall length of an **Airbus A380** (72.7 m) is
    10 × the width of a (soccer) **football goal** (7.32 m)

The **Trans-Canada Highway** (7820 km) is
    twice (2 ×) the length of **Route 66** from Chicago to LA (3940 km)

The length of **Manhattan Island** (21.6 km) is
    400,000 × the height of a **credit card** (54 mm)

BACK OF THE BOOK

Introduction

Which of these is the most numerous?
- ☐ Number of Boeing 747s built (up to 2016): 1520
- ☐ Population of the Falkland Islands: 2840
- ☐ Number of grains of sugar in a teaspoon: 4000
- ☑ **Number of satellites orbiting the Earth (in 2015): 4080**

Web link: http://IsThatABigNumber.com/link/q-intro

Reference

Dehaene, Stanislas, *The Number Sense: How the Mind Creates Mathematics*. Oxford University Press, 1997 (revised and updated edition, 2011)

Worthwhile web links[1]

Carl Sagan explains how Eratosthenes calculated the size of the world
http://IsThatABigNumber.com/link/b-intro-sagan

Samuel Pepys begins his diary
http://IsThatABigNumber.com/link/b-intro-pepys

---

[1] All the links given in this 'Back of the Book' section are reachable from http://IsThatABigNumber.com/link/book

What Counts?

Which of these is the most numerous?
- ☐ Number of aircraft carriers in the world: 167 carriers
- ☑ **Number of skyscrapers in New York City: 250 buildings**
- ☐ Estimated population of Sumatran rhino: 100 individuals
- ☐ Number of bones in the human body: 206 bones

Web link: http://IsThatABigNumber.com/link/q-count

References

Feynman, Richard, *What Do You Care What Other People Think?* W.W. Norton, 1988

Worthwhile web links

US Census Bureau population clock
http://IsThatABigNumber.com/link/b-count-census

Feynman on thinking
http://IsThatABigNumber.com/link/b-count-feynman

Article on counting fish in the sea from International Council for Exploration of the Seas
http://IsThatABigNumber.com/link/b-count-fish1

Estimate of fish biomass
http://IsThatABigNumber.com/link/b-count-fish3

Online subitising test at cognitivefun.net
http://IsThatABigNumber.com/link/b-count-sub1

Guardian article on Daniel Tammet, an autistic savant
http://IsThatABigNumber.com/link/b-count-tammet

Numbers in the World

Which of these weighs the least?
- ☐ A medium-sized pineapple: 900 g
- ☐ A typical pair of men's dress shoes: 860 g
- ☑ **A cup of coffee (cup included): 765 g**
- ☐ A bottle of champagne: 1.6 kg

Web link: http://IsThatABigNumber.com/link/q-numeracy

Worthwhile web links

99% Invisible podcast: The Two Fates of the Old East Portico
http://IsThatABigNumber.com/link/b-numeracy-99pi

ResearchGate: Some acoustical properties of St Paul's Cathedral, London
http://IsThatABigNumber.com/link/b-numeracy-StPauls

About the size of it

Which of these is the longest?

☐ A London bus: 11.23 m

☐ Estimated length of Tyrannosaurus Rex: 12.3 m

☑ **Distance a kangaroo can jump: 13.5 m**

☐ A T-65 X-Wing starfighter in *Star Wars*: 12.5 m

Web link: http://IsThatABigNumber.com/link/q-length

Worthwhile web links

Aztec anthropic units
http://IsThatABigNumber.com/link/b-measure-aztec

Kevin F's project to build the Empire State Building in Lego
http://IsThatABigNumber.com/link/b-measure-legoesb

All about the scale used for Lego minifigs
http://IsThatABigNumber.com/link/b-measure-legoscale

The Spartathon running race
http://IsThatABigNumber.com/link/b-measure-spartathon

Ticking Away

Which of these is the longest period of time?

☑ **Time since the emergence of flowering plants: 125 million years**

☐ Time since the earliest primates: 75 million years

☐ Time since the extinction of dinosaurs: 66 million years

☐ Age of earliest mammoth fossils: 4.8 million years

Web link: http://IsThatABigNumber.com/link/q-time

References

Hofstadter, Douglas, *Metamagical Themas*. Basic Books, 1985
Schofield & Sims, *World History Timeline* [Wall Chart]. Schofield & Sims, 2016

Worthwhile web links

Vox article on Antikythera mechanism
   http://IsThatABigNumber.com/link/b-time-antikythera

Swatch Internet time in .beats
   http://IsThatABigNumber.com/link/b-time-beats

Multidimensional Measures

Which of these has the least volume?

☑ **Volume of water in the Fort Peck Dam (USA): 23 km³**
☐ Volume of water in Lake Geneva: 89 km³
☐ Volume of water in the Guri Dam (Venezuela): 135 km³
☐ Volume of water in the Ataturk Dam (Turkey): 48.7 km³

Web link: http://IsThatABigNumber.com/link/q-volume

Reference

Klein, H. Arthur, *The World of Measurements*. Simon & Schuster, 1974

Worthwhile web links

NewGeography.com article on urban density
   http://IsThatABigNumber.com/link/b-area-urban

Discussion on oil reserves
   http://IsThatABigNumber.com/link/b-volume-oilreserves

BP article on oil tankers
   http://IsThatABigNumber.com/link/b-volume-oiltanker

Metrics on rainfall
   http://IsThatABigNumber.com/link/b-volume-rain

Massive Numbers

Which of these has the greatest mass?

☑ **An Airbus A380 airliner (maximum takeoff weight): 575,000 kg**
☐ The Statue of Liberty: 201,400 kg
☐ An M1 Abrams tank: 62,000 kg
☐ The International Space Station: 420,000 kg

Web link: http://IsThatABigNumber.com/link/q-mass

Worthwhile web links

Article on the Curtiss Helldiver aeroplane
http://IsThatABigNumber.com/link/b-mass-curtiss

Live Science article on Sherpa Guides
http://IsThatABigNumber.com/link/b-mass-sherpa

National Physical Laboratory article on weighing
http://IsThatABigNumber.com/link/b-mass-weighing

Getting up to Speed

Which of these is the fastest?
- ☐ Top speed attained by a human-powered aircraft: 44.3 km/h
- ☑ **Top speed of a giraffe: 52 km/h**
- ☐ Top speed attained by a human-powered watercraft: 34.3 km/h
- ☐ Top speed of a Great White shark: 40 km/h

Web link: http://IsThatABigNumber.com/link/q-speed

Worthwhile web link

Brit Lab video: can a falling penny kill you?
http://IsThatABigNumber.com/link/b-speed-penny

Intermission

Worthwhile web links

Danger levels for sound volume
http://IsThatABigNumber.com/link/b-logs-decibels

Musical note frequencies
http://IsThatABigNumber.com/link/b-logs-keyboard

Wikiversity article on Moore's law and Intel processors
http://IsThatABigNumber.comlink/b-logs-moore1

Data on Moore's law
http://IsThatABigNumber.com/link/b-logs-moore2

UK Office of National Statistics mortality tables
http://IsThatABigNumber.com/link/b-logs-mortality

Antiquark: Online slide rule simulation
http://IsThatABigNumber.com/link/b-logs-sliderule

Heavens Above

Which of these is the biggest?

- ☐ An Astronomical Unit (AU): 149.6 million km
- ☐ Distance from the Sun to Neptune: 4.5 billion km
- ☐ Length (circumference) of Earth's orbit around the Sun: 940 million km
- ☑ **Halley's comet's furthest distance from the Sun (aphelion): 5.25 billion km**

Web link: http://IsThatABigNumber.com/link/q-astro

Worthwhile web links

Powers of Ten: video using log-scale logic to show the scale of the universe
http://IsThatABigNumber.com/link/b-astro-universe

The Cavendish experiment to weigh the world
http://IsThatABigNumber.com/link/b-astro-cavendish

SkyMarvels: video showing the dance of the Earth and the Moon about their barycentre
http://IsThatABigNumber.com/link/b-astro-barycentre

A Bundle of Energy

Which of these is the biggest?

- ☐ Energy released by metabolising 1 gram of fat: 38 kJ
- ☑ **Energy in 1 gram of a meteor hitting the Earth: 500 kJ**
- ☐ Energy released by burning 1 gram of petrol (gasoline): 45 kJ
- ☐ Energy released by exploding 1 gram of TNT: 4.2 kJ

Web link: http://IsThatABigNumber.com/link/q-energy

Worthwhile web links

List of energy densities
http://IsThatABigNumber.com/link/b-energy-density

Blacksmiths' guide to colours of glowing steel
http://IsThatABigNumber.com/link/b-energy-glow

Ignition points
http://IsThatABigNumber.com/link/b-energy-ignite

Our World In Data guide to energy production
http://IsThatABigNumber.com/link/b-energy-production

Sandia Analysis of potential of solar energy
http://IsThatABigNumber.com/link/b-energy-solar2

Worldwide energy consumption
http://IsThatABigNumber.com/link/b-energy-worldwide

Bits, Bytes, and Words

Which of these had the most computer memory?

☐ First Apple Macintosh Computer: 128 KB

☑ **First IBM Personal Computer: 256 KB** maximum

☐ BBC Micro Model B Computer: 32 KB

☐ First Commodore 64 Computer: 64 KB

Web link: http://IsThatABigNumber.com/link/q-info

Reference

Gitt, Werner, *In the Beginning Was Information.* Master Books, 2006

Worthwhile web links

Mashable explanation of Google's enumeration of all the books in the world
http://IsThatABigNumber.com/link/b-info-books

File Catalyst: media file sizes
http://IsThatABigNumber.com/link/b-info-media

How much data does NSA's data centre hold?
http://IsThatABigNumber.com/link/b-info-nsa-data

Zachary Booth Simpson: Vocabulary Analysis of Project Gutenberg
http://IsThatABigNumber.com/link/b-info-vocab

Commonplacebook.com: word count for famous novels
http://IsThatABigNumber.com/link/b-info-words

Let Me Count the Ways

Which of these is the biggest number?
- ☐ **Possible starting hands in Hold 'Em Poker (two cards dealt): 1326**
- ☐ Ways a travelling salesman can visit six towns (and return home): 360
- ☐ Digits used in writing a googol in binary: 333
- ☐ Ways of seating six people around a table: 120

Web link: http://IsThatABigNumber.com/link/q-maths

Worthwhile web links

Wait But Why: A great attempt at explaining the construction of Graham's number
http://IsThatABigNumber.com/link/b-math-grahams

Optimap website will solve the travelling salesman problem for a small number of nodes
http://IsThatABigNumber.com/link/b-math-tsp

Numberphile (a very worthwhile series of YouTube videos) on Infinity
http://IsThatABigNumber.com/link/b-math-infinity

Who Wants to be a Millionaire?

Which of these is the greatest amount?
- ☐ Cost of *Apollo* moon programme (in 2016 dollars): $146 billion
- ☑ **GDP of Kuwait in 2016: $301.1 billion**
- ☐ Turnover of Apple in 2016: $215.6 billion
- ☐ Value of Russia's gold reserves (July 2016): $64.7 billion

Web link: http://IsThatABigNumber.com/link/q-money

Reference

CIA, *World Factbook*. US Directorate of Intelligence, 2017

Worthwhile web links

OECD economic data
http://IsThatABigNumber.com/link/b-money-oecd

Research and development spending by country
http://IsThatABigNumber.com/link/b-money-rnd

Bank of England: UK Inflation calculator
http://IsThatABigNumber.com/link/b-money-uk-inflation1

US Treasury: stats on US national debt
http://IsThatABigNumber.com/link/b-money-us-debt-stats
US Inflation calculator
http://IsThatABigNumber.com/link/b-money-us-inflation

Everybody Counts

Which of these is the greatest number?
- ☐ Population of Chongqing, China: 8.19 million
- ☑ **Population of Austria: 8.71 million**
- ☐ Estimated world population of blue duiker antelope: 7 million
- ☐ Population of Bulgaria: 7.14 million

Web link: http://IsThatABigNumber.com/link/q-pop

Reference

Brunner, John, *Stand on Zanzibar*. Doubleday, 1968

Worthwhile web links

Index Mundi: Protected areas over time
http://IsThatABigNumber.com/link/b-population-indexmundi
World Population History: Map-based presentation
http://IsThatABigNumber.com/link/b-population-map
United Nations: World population prospects report
http://IsThatABigNumber.com/link/b-population-un
World Bank: Map of protected areas
http://IsThatABigNumber.com/link/b-population-worldbank
The Millions: Essay on John Brunner's *Stand on Zanzibar*
http://IsThatABigNumber.com/link/b-population-zanzibar

Measuring How We Live

Which of these countries scores highest for happiness?
- ☐ United Kingdom: 6.714
- ☐ Canada: 7.316
- ☐ Costa Rica: 7.079
- ☑ **Iceland: 7.504**

Web link: http://IsThatABigNumber.com/link/q-quality

References

United Nations, *Millennium Development Goals Report.* United Nations, 2015

Helliwell, J., Layard, R. & Sachs, J., *World Happiness Report 2016, Update* (Vol. I). Sustainable Development Solutions Network, 2016

Worthwhile web links

The Stealth ride at Thorpe Park, UK
  http://IsThatABigNumber.com/link/b-quality-stealth

The outstanding 'Our World In Data' website on inequality
  http://IsThatABigNumber.com/link/b-quality-owid

Summing Up

Which of these is most true?
☐ The end is nigh
☐ It was better in the old days
☐ We'll muddle through
☐ We're heading for the sunlit uplands

You'll have to make up your own mind on this one.

Worthwhile sources of numbers

Our World In Data
  http://IsThatABigNumber.com/link/b-summary-owid

CIA Factbook
  http://IsThatABigNumber.com/link/b-summary-cia

Gapminder
  http://IsThatABigNumber.com/link/b-summary-gm

World Bank
  http://IsThatABigNumber.com/link/b-summary-wb

OECD
  http://IsThatABigNumber.com/link/b-summary-oecd

ONS
  http://IsThatABigNumber.com/link/b-summary-ons

XKCD
  http://IsThatABigNumber.com/link/b-summary-xkcd

ACKNOWLEDGEMENTS

Thanks are due to everyone who encouraged me in writing this book. Most especially to my wife, Beverley Moss-Morris, without whose encouragement and support this would have been abandoned on a hundred occasions.

Thanks to my sons who from an early age challenged my thinking, keeping it sharp and, I hope, honest.

Thanks to my sister, Rose Malinaric, the invisible editorial hand whose keen eye and sharp judgement helped so much in knocking successive drafts into shape.

Thanks to my agent Leslie Gardner, of the Artellus agency, who decided that a book on numbers might have an audience, and found just the right publisher for it.

Thanks to all at Oxford University Press and particularly Dan Taber for his confidence in the project. Thanks, too, to all those involved in the editing and preparation of the final text, and especially to Mac Clarke, who lent support when it mattered. Thanks also to Lisa Eaton, UK Project Manager at SPi Global, for her great patience with an inexperienced author.

Thanks to all my friends, but especially Mark and Erika Cropper for their unfailing friendship and their enthusiasm for this project, and their belief in me. Thanks to everyone who has over the years listened patiently to me discussing my oddball passions and enthusiasms.

There are so many facts and figures contained in this book that it would be extraordinary if some mistakes had not slipped through. Of course, I take responsibility for any such errors. For anyone spotting a number-related error, though, I say: "Well done, you've clearly been exercising your numerical skills to good effect!"

Andrew C. A. Elliott
Peper Harow
February 2018

INDEX

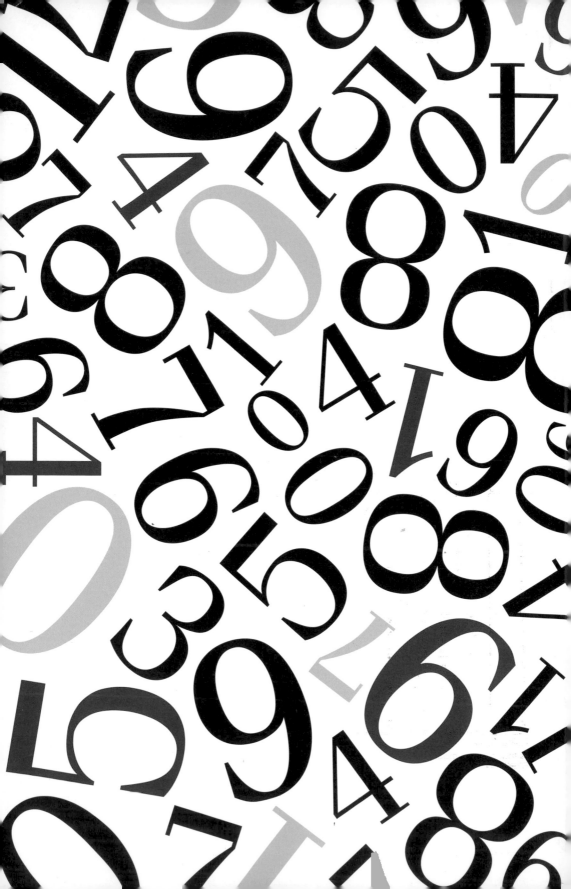